CLIMATE CHANGE
THE UK PROGRAMME

United Kingdom's Report under the Framework Convention on Climate Change

Presented to Parliament by the Secretaries of State for the Environment and the Foreign and Commonwealth Office, the Chancellor of the Exchequer, the President of the Board of Trade, the Secretaries of State for Transport, Defence, National Heritage and Employment, the Chancellor of the Duchy of Lancaster, the Secretaries of State for Scotland, Northern Ireland, Education and Health, the Minister for Agriculture, Fisheries and Food, the Secretary of State for Wales and the Minister for Overseas Development by Command of Her Majesty.
January 1994.

Cm 2427 LONDON: HMSO £10.00 net

LIVERPOOL INSTITUTE OF HIGHER EDUCATION	
Order No./Invoice No.	£10.84 L3278/B698030
Accession No.	150838
Class No.	QUARTO 910.515 CLI
Control No.	ISBN
Catal.	4 6/94

FOREWORD

BY THE PRIME MINISTER

At the Earth Summit in Rio in June 1992, I signed the Framework Convention on Climate Change, on behalf of the United Kingdom. 150 other countries also signed, and more have done so since. In doing so, each of us recognised the daunting challenge posed by the threat of climate change, and the urgent need for precautionary action to be taken.

In order to help encourage a prompt start to action, I pledged at the time that the UK would produce a programme of measures to implement the Convention's commitments as quickly as possible. This programme fulfils that pledge. It sets out the measures we are taking to meet all our commitments, notably those aimed at returning emissions of carbon dioxide and each other major greenhouse gas to 1990 levels by 2000.

The threat of climate change cannot be tackled by any one country or group of countries in isolation; it requires a global response. The Convention we signed at Rio represents an important first step towards providing the framework for such a response. At the same time, the measures required to tackle the problem cannot simply be decreed by governments; they depend on decisions and actions taken by all individual citizens in their daily lives. For this reason, this programme was drawn up after widespread consultation within the United Kingdom. It seeks to harness the efforts of all sectors of our society and the participation of all our people in meeting the challenge for the sake of future generations.

John Major

SUMMARY

The Climate Change Programme is the UK's first report under Article 12 of the Framework Convention on Climate Change, which the UK ratified in December 1993.

The UK's programme reflects the commitment to adopt a precautionary approach (see Chapter One). In other words, the threat of climate change is sufficiently great to require action now, even though scientific uncertainties remain. This is particularly appropriate where low cost preventive action can be taken at an early stage to avoid more expensive action later. The programme takes account of the UK's particular national circumstances (see Chapter Two).

The programme includes inventories of emissions and scenarios of possible future trends, and sets out measures aimed at returning emissions of each of the main greenhouse gases to 1990 levels by 2000. It comprises:

- for carbon dioxide (CO_2), a programme based on a national partnership approach (see Chapter Three) drawn up after extensive public consultation involving business, voluntary, environmental and other interest groups. The result is a comprehensive programme of measures aimed at saving about 10 million tonnes of carbon (MtC) against projected emissions by 2000, spread across all sectors:
  - domestic energy consumption (**4MtC**), through measures including introducing taxation on domestic fuel and power, and a new Energy Saving Trust;
  - energy consumption by business (**2.5MtC**), including through energy efficiency advice and information;
  - energy consumption by the public sector (**1MtC**), through targets for central and local government and public sector bodies; and
  - transport (**2.5MtC**), through increases in road fuel duties, and a commitment to real increases of at least 5% on average in future Budgets.
- for methane (see Chapter Five), an anticipated 10% reduction from 1990 levels by 2000, including through initiatives to encourage the use of methane for energy generation, such as new waste management practices, and reduced emissions from coal production. New guidance on limiting emissions from several industrial sectors is also being prepared;
- for nitrous oxide (see Chapter Six), an anticipated 75% reduction from 1990 levels by 2000, mainly through the reduction of nitrous oxide emissions from nylon manufacture; and
- major anticipated reductions from 1990 levels by 2000 for other greenhouse gases (see Chapter Six), including nitrogen oxides (a 25% fall), volatile organic compounds (a 35% fall), carbon monoxide (a 50% fall), halocarbons, such as chlorofluorocarbons (elimination), and perfluorocarbons, such as carbon tetrafluoride (a 90% fall) and hexafluoroethane (a 90% fall).

Chapter Four sets out how the UK is meeting its commitment on enhancing carbon sinks in soils and vegetation. Estimates suggest that there was a net uptake of between 1.5 and 2.5 MtC in 1990. Afforestation and other new tree planting are estimated to be fixing carbon at a rate of 2.5MtC a year.

The UK is providing substantial assistance to developing countries that will help combat climate change (see Chapter Seven) through bilateral aid, particularly for energy efficiency and forestry projects, and through multilateral channels, including the Global Environment Facility. The UK also provides multilateral and bilateral help to Eastern European and other countries with economies in transition to combat climate change.

The programme includes (see Chapter Eight) a wide range of publicity and educational campaigns to improve awareness of climate change.

The UK's research effort places it at the forefront of climate change research (see Chapter Nine). The UK spent over £140 million on climate change research in 1992/93 including work on data acquisition, climate prediction and modelling, particularly at the Hadley Centre, response measures and impacts and adaptation. The UK gives strong support to the work of the Intergovernmental Panel on Climate Change, and chairs its science working group.

The Government recognises that we need to plan ahead now for further actions that may be needed beyond 2000. Chapter Ten assesses possible longer term developments and options.

CONTENTS

CHAPTER		PAGE
1	INTRODUCTION	6
2	THE UK PROGRAMME IN CONTEXT	8
3	THE GREENHOUSE GAS PROGRAMME – CARBON DIOXIDE (CO_2)	14
	• CO_2 emissions inventory and trends	
	• Measures to limit emissions	
	– Electricity generation and supply	
	– Energy consumption in the home	
	– Energy consumption by business	
	– Energy consumption in the public sector	
	– Transport	
	• Costs and economic impact	
	• Monitoring	
4	THE GREENHOUSE GAS PROGRAMME – CARBON RESERVOIRS AND SINKS	35
5	THE GREENHOUSE GAS PROGRAMME – METHANE	39
	• Methane emissions inventory and trends	
	• Measures to limit emissions	
	• Monitoring	
6	THE GREENHOUSE GAS PROGRAMME – OTHER GREENHOUSE GASES	47
	• Nitrous oxide	
	• Tropospheric ozone	
	• Nitrogen oxides	
	• Volatile organic compounds	
	• Carbon monoxide	
	• Halocarbons	
	• Perfluorocarbons	
7	ASSISTANCE TO DEVELOPING COUNTRIES AND COUNTRIES WITH ECONOMIES IN TRANSITION	54
8	EDUCATION, TRAINING AND PUBLIC AWARENESS	59
9	UK CLIMATE CHANGE RESEARCH	63
	• UK contribution to international research	
	• UK research programme	
	– Data for climate change research	
	– Climate prediction and monitoring	
	– Response measures	
	– Impacts and adaptation	
10	BEYOND 2000: TOWARDS SUSTAINABILITY	68
	• Climate change beyond 2000	
	• The UK contribution	
	– Transport	
	– Electricity generation	
	– Energy consumption	
	– Afforestation	
ANNEX A	LIST OF ABBREVIATIONS	73
ANNEX B	UK NATIONAL ATMOSPHERIC EMISSIONS INVENTORY	76

CHAPTER 1

INTRODUCTION

> "In order to protect the environment the precautionary approach shall be adopted by states according to their capabilities. Where there are threats of serious damage, lack of full scientific certainty shall not be used as a reason for postponing cost-effective measures to prevent environmental degradation."
>
> **Principle 15, Rio Declaration on Environment and Development**

[handwritten margin note: ✱ Published twelve years ago need more recent figures (Ref to new Kyoto Policy)]

1.1 This report sets out the United Kingdom's programme for meeting the commitments in the United Nations Framework Convention on Climate Change (the Convention)[1]. It constitutes the UK's first report to the Conference of the Parties, as required under Article 12 of the Convention. It has been produced ahead of the schedule in Article 12.5 to demonstrate the importance the UK attaches to the Convention and to making a prompt start on its implementation.

1.2 The UK signed the Convention in June 1992 and ratified it in December 1993[2]. As a developed country party, the UK accepts the particular obligations this entails, including the commitment to take measures aimed at returning emissions of greenhouse gases to 1990 levels by the year 2000. The UK has therefore prepared a detailed programme of measures designed to achieve this commitment for each of the main greenhouse gases and to fulfil the other commitments in the Convention, including those on assistance to developing countries, on protecting and enhancing carbon sinks (such as forests), on supporting research into climate change, and on promoting public education and awareness. The report sets out the measures that the UK Government is committed to undertake in each of these areas.

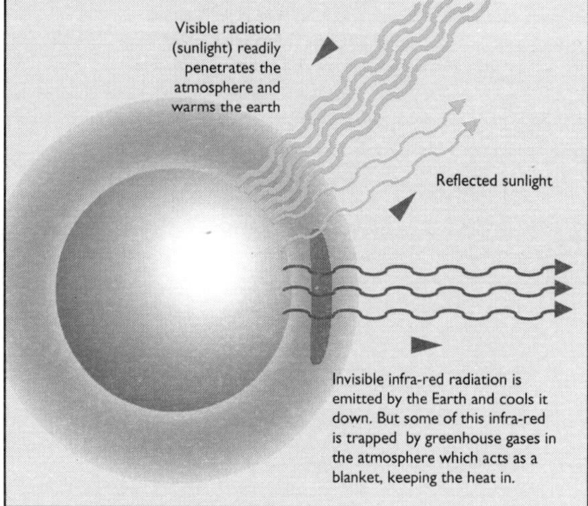

1.3 The centrepiece of the programme is the set of measures designed to limit emissions of carbon dioxide (CO_2), the most important greenhouse gas. This is set out in Chapter Three. The UK's CO_2 Programme is based on a partnership approach and was developed following extensive consultation. This reflects the fact that decisions that help to limit CO_2 emissions are ultimately made not only by governments, but also by individuals – businessmen and businesswomen, those at work, homeowners, transport users, anyone who uses energy in all its forms.

PARTNERSHIP APPROACH

1.4 The Government has therefore been concerned to encourage commitment to the UK's national programme by seeking public participation in drawing it up. Consumer groups, environmental, housing and transport organisations, local authorities, business organisations, energy utilities and central government have all contributed to the decision-making process. Ten thousand copies of a discussion document[3] were issued in December 1992 to encourage a major debate on the UK's national programme, before final decisions were taken. The ensuing debate was the first of its kind in the UK, examining the part individuals, businesses and the public sector could play in limiting CO_2 emissions, and the options for Government measures to support such action.

1.5 A series of workshops to look at different sectors of the economy was held in March 1993. This was followed by a major national conference in May 1993 at which the workshop findings were discussed by the leaders of business, industry, consumer and environmental groups. A clear view emerged from the consultation process that the Government must provide a lead and set the basic legislative and fiscal framework for the national programme. This would provide the right conditions within which others could act. During the consultation process, many organisations expressed a willingness to play their part by taking action to improve their own energy and fuel efficiency, and to help others to do so. This is the essence of the partnership approach.

PRECAUTIONARY APPROACH

1.6 The UK's programme is based around the precautionary approach inherent in the Convention. The work of the Intergovernmental Panel on Climate Change (IPCC) has shown that the threat of climate change is such that it is appropriate to take action ahead of unequivocal evidence being established about the nature and possible effects of

Figure 1a The greenhouse effect

Visible radiation (sunlight) readily penetrates the atmosphere and warms the earth

Reflected sunlight

Invisible infra-red radiation is emitted by the Earth and cools it down. But some of this infra-red is trapped by greenhouse gases in the atmosphere which acts as a blanket, keeping the heat in.

EVIDENCE FROM IPCC ON CLIMATE CHANGE AND ITS EFFECTS

The UK has been a leading supporter of, and contributor to, the scientific work of the Intergovernmental Panel on Climate Change (IPCC) – see Chapter Nine. The IPCC produced its First Assessment Report[4] in 1990, representing the views of over 300 leading scientists from around the world. Its major conclusions were that:

- the concentration of greenhouse gases in the atmosphere has increased substantially as a result of human activity;
- this is expected to enhance the natural greenhouse effect, which keeps the earth warmer than it would otherwise be;
- average global temperatures have increased by 0.3°C to 0.6°C during the last century. This is consistent with models of the enhanced greenhouse effect that would be expected, but it is also within the bounds of normal climate variability;
- without actions to restrain emissions, an increase in global average temperatures of around 0.3°C per decade (with an uncertainty range of 0.2 to 0.5°C) is likely in the future. This could imply sea level rise of around 6 cm per decade (with an uncertainty range of 3 to 10 cm) (see paragraph 9.17 on impacts).

These findings were confirmed in the IPCC's 1992 supplementary report[5], which also produced some new understandings of the complex processes involved. Both reports recognise that uncertainties remain. It will be some years before the warming effect moves beyond the bounds of natural climate variability. More work is needed on improving understanding of the climate processes involved to improve the predictive accuracy of the models, particularly at the regional level. The IPCC's Second Assessment Report, due in 1995, will address these and other issues. Meanwhile, observational data continues to confirm the recent trend: globally, 1990 was the warmest year since records began in about 1860; and all but one of the warmest 11 years on record have occurred since 1980 (see Figure 1b).

man-made climate change (see box on IPCC). Such an approach is particularly appropriate where low cost preventive action can be taken at an early stage to avoid more expensive action later. The UK programme takes advantage of the considerable scope that exists for taking cost-effective action now. It also emphasises the importance of exploring the full range of policy tools available, including economic instruments, regulatory measures, voluntary action and public information.

1.7 The UK recognises that the requirements of the current framework Convention may well represent only a first step, though an important one, towards the achievement of its ultimate objective, as set out in Article 2. The final Chapter of this report describes some of the policy options that the UK Government will need to consider in the context of the further evolution of the Convention, and the scientific evidence underpinning it. For the present, however, the priority for the UK, and for all nations, must be to set out clearly and comprehensively how they will meet their obligations under the Convention as it stands. That is what this report aims to do.

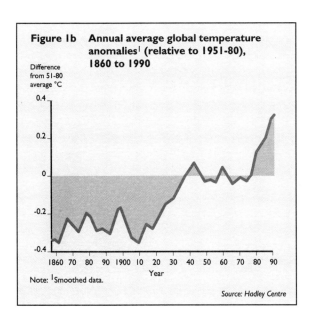

Figure 1b Annual average global temperature anomalies[1] (relative to 1951-80), 1860 to 1990

Note: [1] Smoothed data.

Source: Hadley Centre

NOTES

1. Report of the Intergovernmental Negotiating Committee for a Framework Convention on Climate Change on the Work of the Second part of its Fifth Session, held at New York from 30 April to 9 May, 1992.
2. The UK is encouraging its Dependent Territories, where appropriate, to be included in the UK's ratification of the Convention. The Government is discussing with them what they would need to do to play their part in the efforts made by the UK to implement the Convention's requirements.
3. Climate Change: Our National Programme for CO_2 Emissions – a discussion document. The Department of the Environment, 1992.
4. Intergovernmental Panel on Climate Change First Assessment Report 1990. Cambridge University Press, 1990. ISBN 0-52-140720-6.
5. Climate Change 1992: The Supplementary Report to the IPCC Scientific Assessment, Houghton J T, Callender B A and Varney S K (eds). Cambridge University Press, 1992.

CHAPTER 2

THE UK PROGRAMME IN CONTEXT

2.1 The Convention recognises in Article 4.2a that action needs to take account of differences in "starting points and approaches, economic structures and resource bases". The UK programme has been prepared taking account of our particular national circumstances.

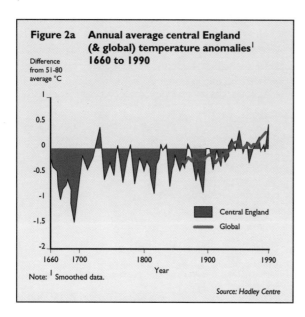

Figure 2a Annual average central England (& global) temperature anomalies[1] 1660 to 1990

Note: [1] Smoothed data.

Source: Hadley Centre

CLIMATE

2.2 The UK's climate is variably cool, moist, temperate and maritime with a moderate annual average temperature (see Figure 2a) and limited ranges. Space heating is needed in buildings throughout the winter months, but there is little need for additional air conditioning in the summer. Nevertheless, air conditioning is increasingly used in offices and factories.

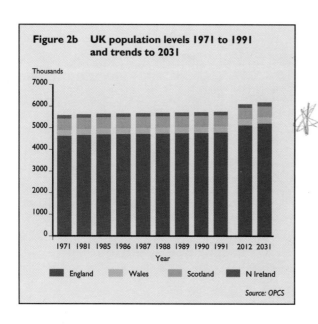

Figure 2b UK population levels 1971 to 1991 and trends to 2031

Source: OPCS

POPULATION

2.3 The UK's population of 57.6 million represents growth of 3% over the last twenty years. It is expected to grow by 6% over the next twenty (see Figure 2b). The UK is experiencing a reduction in the average number of people in households, which will result in a 15% rise in the number of households over the next two decades. These demographic trends are likely to exert gradually increasing pressure on resources and energy supplies into the next century. 47% of the population is concentrated in the English South East and Midlands, which are among the most densely populated areas of the world, with consequent implications for patterns of demand for energy and other resources. In contrast, other areas, including the Scottish Highlands, parts of Wales and the North and South West of England, are sparsely populated.

LAND USE

2.4 The land area of the UK is 24.3 million hectares, with all but a very small proportion devoted to some form of economic activity. Land which is not used for agriculture or forestry, the major part of which is land in urban use, accounts for around three and a half million hectares. Agriculture is an important economic sector, and accounts for around 18.5 million hectares of land and 32% of the UK's methane emissions. Land used for agriculture has declined slightly over the last 20 years, and that used for forest and woodland and urban land has increased. Only 10% of land is forested, although this total is expanding under the influence of Government programmes. Existing forests absorb about 2.5 million tonnes of carbon (MtC) annually (see Chapter Four). The UK's household, industrial and commercial sectors produce about 140 million tonnes of "controlled waste" annually, of which 70% is disposed of in landfill sites. While these sites account for a small fraction of the UK's land surface, they produce 39% of the UK's methane emissions.

ENERGY

2.5 The UK is a major producer of natural gas and coal. It is a producer and exporter of oil. These fuels, together with nuclear power, are the UK's primary sources of energy. Renewable energy sources, including hydroelectricity, currently account for a small proportion of electricity and heat generation.

2.6 Figures 2c and 2d show UK primary energy consumption in 1990.

2.7 The energy intensity of the UK economy (as measured by the ratio of primary energy consumption and Gross Domestic Product) has been on a declining path for over one hundred

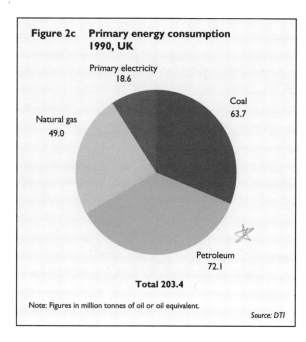

Figure 2c   Primary energy consumption 1990, UK

- Primary electricity 18.6
- Coal 63.7
- Natural gas 49.0
- Petroleum 72.1

Total 203.4

Note: Figures in million tonnes of oil or oil equivalent.

Source: DTI

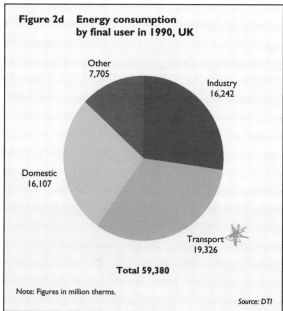

Figure 2d   Energy consumption by final user in 1990, UK

- Other 7,705
- Industry 16,242
- Domestic 16,107
- Transport 19,326

Total 59,380

Note: Figures in million therms.

Source: DTI

years, with the rate of decline tending to speed up in periods of relatively high energy prices and slow down when energy prices were lower[1] (see Figures 2e and 2f). Between 1990 and 1992 the energy intensity rose slightly but indications are that in 1993 it has reverted to its downward path as the UK economy moves out of recession. Over time the rise in energy consumption has generally been at a slower rate than the rise in GDP. Between 1970 and 1990, the UK's GDP rose by 54% in real terms, whilst primary energy consumption increased by only 6%. The trend towards lower intensity has been due to structural changes in the economy, technology changes, and the greater efficiency with which energy is converted, distributed and used.

2.8 In the industrial sector, between 1970 and 1990 final energy consumption per unit of manufacturing output fell by 41%. It is not easy to separate out the relative impact of energy efficiency improvements, and structural and technology changes, but evidence suggests that improvements in energy efficiency may have accounted for at least half of the reduction in energy intensity over this period[2]. Although much of the industrial sector has updated its technology and plant, there is still scope for increasing efficiency. Similarly, in other sectors much could still be done to improve energy efficiency.

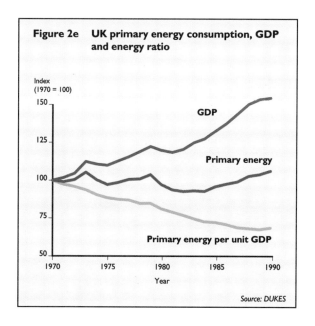

Figure 2e   UK primary energy consumption, GDP and energy ratio

Source: DUKES

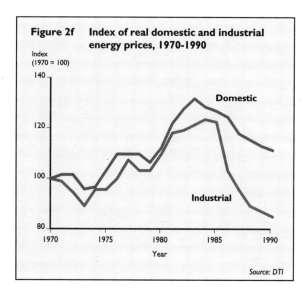

Figure 2f   Index of real domestic and industrial energy prices, 1970-1990

Source: DTI

2.9 In the diverse services sector, which includes shops, offices and schools, energy intensity measured in terms of primary energy per unit of service output declined by 20% over the same period, partly because of efficiency improvements to heating appliances and lighting systems, and also some gains due to fuel switching.

2.10 Domestic energy efficiency has increased by one third, or approaching 2% per year, over the last 20 years. Recent research suggests that energy consumption in 1991 would have been 50% higher if standards of wall, loft and tank insulation and efficiencies of heating appliances were the same as those in 1970[3]. The widespread switches from solid fuel to gas and from non-central to central heating have also played a prominent part.

2.11 The carbon intensity of primary energy consumption in the UK fell by some 18% between 1970 and 1990 as oil, natural gas and nuclear energy replaced coal. The rise in the use of natural gas and the increasing use of nuclear energy were particularly significant. In some sectors the greater importance of electricity as a fuel has tended to have an offsetting effect on the fall in the average carbon intensity, when considering energy delivered to final users. This effect is notable in industry and especially in the commercial/public services sector, where the average carbon intensity has risen. Figure 2g shows the overall effect on carbon intensity per unit delivered energy when power station (and refinery) emissions are shared amongst final users in the four main sectors of the economy.

2.12 The range of scenarios used in the CO_2 programme in Chapter Three indicates a fall in the carbon intensity of energy in the first half of the 1990s, partially offset by a slight rise in the second half of the decade as the share of nuclear energy in total use begins to decline. However, the programme aims to reinforce the trend to lower energy intensity by ensuring that a greater proportion of the remaining potential for improvements in energy efficiency is realised in all sectors than might otherwise be achieved.

TRANSPORT

2.13 In 1990, the transport sector was responsible for 24% of UK anthropogenic CO_2 emissions, about 86% of which came from road vehicles (see Figure 2h). CO_2 emissions from the transport sector have increased from 23.4 MtC in 1970 to 38 MtC in 1990 (see Figure 2i). Moreover, the transport sector now represents one of the fastest growing sources of UK CO_2 emissions.

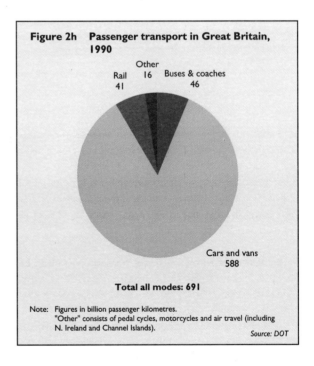

Figure 2h Passenger transport in Great Britain, 1990

Rail 41
Other 16
Buses & coaches 46
Cars and vans 588

Total all modes: 691

Note: Figures in billion passenger kilometres.
"Other" consists of pedal cycles, motorcycles and air travel (including N. Ireland and Channel Islands).

Source: DOT

Figure 2g Carbon to delivered energy ratio for the main sectors of the UK economy 1970-90[1]

MtC/Mtoe

Domestic   Industrial   Public   Transport

Note: [1] Measured as tonnes of carbon per unit delivered energy (expressed in tonnes of oil equivalent).
Power station and refinery emissions have been shared amongst end users.

Source: DUKES, UK-NAEI

2.14 Between 1970 and 1990, fuel consumption by cars rose by around 70%, car ownership almost doubled and car mileage (including taxis) grew by around 117% (see Figure 2j). There has been a close correlation between car ownership, car mileage and economic growth: as the UK economy has developed more people have been able to afford to run a vehicle, and to travel further in the vehicles they own. The efficiency with which cars consume fuel has increased over the last 30 years.

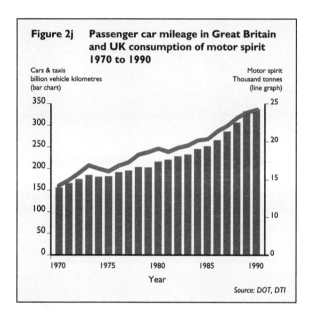

Figure 2j Passenger car mileage in Great Britain and UK consumption of motor spirit 1970 to 1990

Source: DOT, DTI

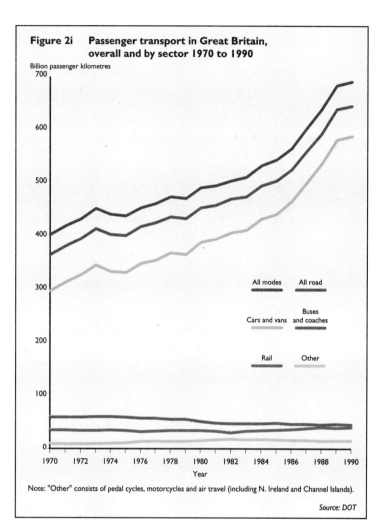

Figure 2i Passenger transport in Great Britain, overall and by sector 1970 to 1990

Note: "Other" consists of pedal cycles, motorcycles and air travel (including N. Ireland and Channel Islands).

Source: DOT

2.15 Public transport has the ability to move large numbers of people much more efficiently than cars but cannot cater easily for the scattered pattern of journeys that has accompanied the growth of car ownership and the dispersal of land use. Public transport's share of the number of people who travel has declined over recent years to 13% of all journeys of more than one mile.

2.16 Industry requires an efficient transport system to be able to function effectively. UK industry moved 2.16 billion tonnes of goods in 1990 (see Figure 2k). Freight haulage by road has shown sustained growth, particularly in manufactured and expensive products. It is expected to grow with the completion of the European internal market.

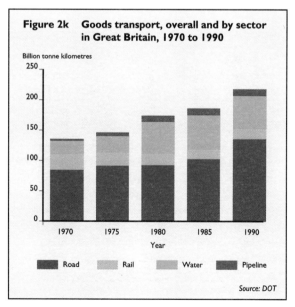

Figure 2k Goods transport, overall and by sector in Great Britain, 1970 to 1990

Source: DOT

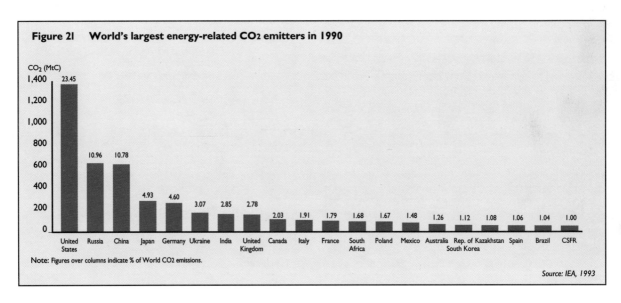

Figure 2l World's largest energy-related CO_2 emitters in 1990

Note: Figures over columns indicate % of World CO2 emissions.

Source: IEA, 1993

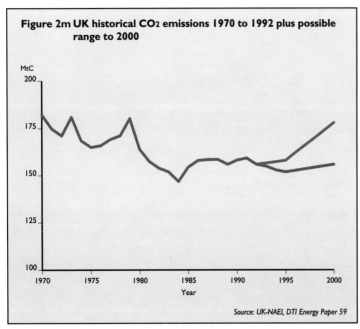

Figure 2m UK historical CO_2 emissions 1970 to 1992 plus possible range to 2000

Source: UK-NAEI, DTI Energy Paper 59

TRENDS IN EMISSIONS OF CARBON DIOXIDE

2.17 Figure 2l shows UK CO_2 emissions compared with other countries in 1990. The Government published a range of projections of energy demand and related CO_2 emissions as Energy Paper 59[4] (EP 59) in October 1992 (see Figure 2m). These projections were derived from economic and statistical analysis of the energy market using current knowledge about likely technological developments in the future. They incorporated historical trend improvements in energy efficiency and the effects of on-going Government policies, but did not assume any impact from additional measures yet to be implemented after the end of 1991.

2.18 Key assumptions which have to be made over the projection period are those for trend GDP growth, fuel prices, the evolution of the structure of the economy and the cost of new technologies. In order to reflect the inevitable uncertainty in these assumptions, the EP 59 range was produced using a number of different trend GDP and fuel price assumptions. The results and the underlying assumptions can then inform debate about longer term trends and the potential impact of policy measures. Projections were made for emissions from the key economic sectors, as well as by fuel.

2.19 Projected CO_2 emissions in EP 59 range under different scenarios from 3 MtC below the initial estimate of 1990 emissions by the year 2000, to 19 MtC above. The range indicates that UK CO_2 emissions could be stable or even falling up to the mid-1990s, mainly due to the greater use of gas and nuclear energy for electricity generation.

TRENDS IN EMISSIONS OF OTHER GREENHOUSE GASES

2.20 After CO_2 the most significant greenhouse gases are methane and nitrous oxide. Figure 2n shows UK greenhouse gas emissions in 1990, with the relative contribution of each gas to global warming. Chapters Five (methane) and Six (other greenhouse gases) set out the 1990 estimates and possible trends to 2000 of UK emissions of these gases. Several of these estimates are subject to a substantial margin of uncertainty and work is in hand to improve them. In many cases the indications of possible trends are based on provisional assumptions which will be revised as estimates are improved.

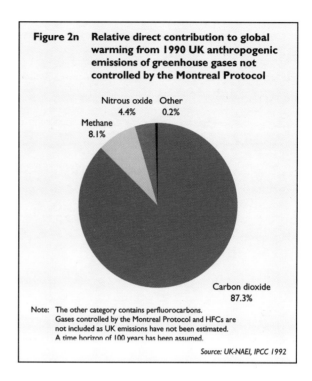

Figure 2n Relative direct contribution to global warming from 1990 UK anthropogenic emissions of greenhouse gases not controlled by the Montreal Protocol

Nitrous oxide 4.4%
Other 0.2%
Methane 8.1%
Carbon dioxide 87.3%

Note: The other category contains perfluorocarbons. Gases controlled by the Montreal Protocol and HFCs are not included as UK emissions have not been estimated. A time horizon of 100 years has been assumed.

Source: UK-NAEI, IPCC 1992

NOTES

[1] *Economic Trends* Number 427 May 1989, Article on "Energy Consumption in the UK".
[2] *Industrial Energy Markets*, Energy Efficiency Series Number 9 (1989).
[3] *Domestic Energy Factfile*, Building Research Establishment (1992).
[4] *Energy Paper 59, Energy Related Carbon Emissions in Possible Future Scenarios for the United Kingdom.* HMSO, 1992. ISBN 0-11-414157-6.

CHAPTER 3

THE GREENHOUSE GAS PROGRAMME – CO_2

INTRODUCTION

3.1 The UK is committed under Articles 4.1 and 4.2 of the Convention to draw up an inventory of greenhouse gas emissions, and to take measures aimed at returning emissions of each greenhouse gas to 1990 levels by 2000. This Chapter describes the programme to meet that commitment for CO_2 emissions, "The CO_2 Programme".

3.2 The CO_2 Programme consists of:
- an inventory of emissions in 1990 (the base year);
- a projection of emissions for the year 2000 (the target year) without abatement measures in place;
- a set of abatement measures aimed at returning emissions to 1990 levels by 2000; and
- a monitoring procedure.

CO_2 EMISSIONS INVENTORY

3.3 The CO_2 Programme takes its estimates for emissions in the base year from the UK National Air Emissions Inventory (UK-NAEI)[1]. The Government publishes the UK-NAEI annually in the Digest of Environmental Protection and Water Statistics[2], which also fulfils the data requirements of other international agreements.

3.4 In order to produce its inventory of CO_2 emissions, the UK-NAEI draws extensively on the Government's annual Digest of UK Energy Statistics (DUKES)[3]. For the energy data used in the calculation of emissions, DUKES is compatible with the UN and International Energy Agency (IEA) systems of international statistics which underpin the emissions methodology being jointly developed by IPCC and the Organisation for Economic Co-operation and Development (OECD). Therefore, the emission inventory from the UK-NAEI should be consistent with that from the IPCC methodology, once final agreement is reached on the source categories to be included there.

3.5 Total UK anthropogenic emissions of CO_2 in 1990 are estimated to have been 158 MtC. This is a reduction of some 2 MtC below our previous estimate due to revisions in the fuel statistics and a reduction in the natural gas emission factor. 1990 was a particularly warm year in the UK – temperatures during the winter months were some 1.7°C above the average for recent years – and economic growth was starting to slow, prior to actual contraction in 1991. These factors had a significant impact on 1990 emission levels – in particular, the high average temperature will have reduced emissions by some 5 MtC below average. The inventory of these emissions can be presented in a number of ways:
- on the basis of the fuel used (see Figure 3a); or

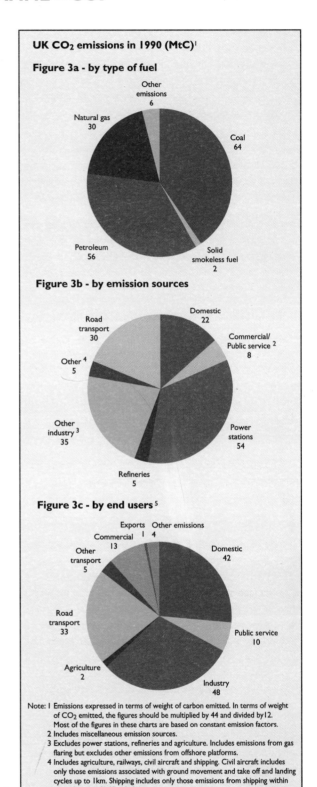

UK CO_2 emissions in 1990 (MtC)[1]

Figure 3a - by type of fuel
- Coal 64
- Natural gas 30
- Other emissions 6
- Petroleum 56
- Solid smokeless fuel 2

Figure 3b - by emission sources
- Domestic 22
- Commercial/Public service[2] 8
- Power stations 54
- Refineries 5
- Other industry[3] 35
- Other[4] 5
- Road transport 30

Figure 3c - by end users[5]
- Domestic 42
- Public service 10
- Industry 48
- Agriculture 2
- Road transport 33
- Other transport 5
- Commercial 13
- Exports 1
- Other emissions 4

Note:
1 Emissions expressed in terms of weight of carbon emitted. In terms of weight of CO_2 emitted, the figures should be multiplied by 44 and divided by 12. Most of the figures in these charts are based on constant emission factors.
2 Includes miscellaneous emission sources.
3 Excludes power stations, refineries and agriculture. Includes emissions from gas flaring but excludes other emissions from offshore platforms.
4 Includes agriculture, railways, civil aircraft and shipping. Civil aircraft includes only those emissions associated with ground movement and take off and landing cycles up to 1km. Shipping includes only those emissions from shipping within coastal waters (< 12 miles).
5 Figures for end users include emissions from power stations which have been allocated, on an approximate basis, to the various sectors according to their use of electricity generated. They also include allocations of emissions from other secondary fuel producers.

Source: HMSO Digest of Environmental and Water Statistics 1992: CO_2 estimated emissions by emission source, end user and type of fuel 1981-1991

- on the basis of the economic sector involved with emissions from fuel processing (ie from refineries or power stations) being shown separately, (see Figure 3b); or
- on the basis of the economic sector with emissions from fuel processing redistributed to end users, (see Figure 3c).

3.6 The treatment of aviation and marine bunkers and the small category of non-fuel related CO_2 emissions may need to be revised once methodologies and definitions are confirmed by the first Conference of Parties to the Convention. These changes are not expected to alter the total estimate of CO_2 emissions by more than a few percentage points. By extension, the CO_2 Programme is likely to be relatively unaffected by any changes in the inventory, but this will be kept under review.

3.7 Estimates of CO_2 emissions rely on the coverage of all emission sources and the accuracy of data on energy consumption and emission factors. Data on energy consumption are compiled from comprehensive statistical systems that collect information on energy supplies to consumers. Such systems provide reliable data of a high quality and are unlikely to lead to major inaccuracies in the measurement of CO_2 emissions.

3.8 Annex B gives further details of the UK-NAEI. A description of the methodology has been published separately.

TRENDS IN EMISSIONS

3.9 Chapter Two (paragraphs 2.17 to 2.19) explains the approach used by the UK in analysing possible future trends in emissions in the absence of any additional abatement measures (see also Figure 3d). Different assumptions about economic growth rates and energy prices are used to provide a view of the possible range for the longer-term development of underlying emissions up to 2000. Although these are not the only assumptions made, they are of particular relevance as both economic growth and fossil fuel prices have a major influence over energy demand and hence CO_2 emissions.

3.10 To provide a focus for the development of a programme of measures to meet the UK's Convention commitment, the Government has selected a representative scenario from the range described in Chapter Two. This "central growth/low fuel price" scenario, now referred to as the 'reference' scenario, is roughly in the centre of the range for the year 2000. Under this scenario emissions are

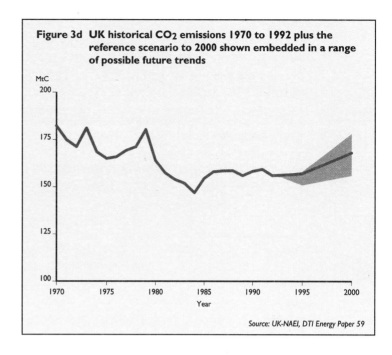

Figure 3d UK historical CO_2 emissions 1970 to 1992 plus the reference scenario to 2000 shown embedded in a range of possible future trends

Source: UK-NAEI, DTI Energy Paper 59

projected to be some 10 MtC higher in 2000 than in 1990. ("Central growth" refers to the assumption that GDP will continue to grow at an average annual rate equal to the historic average peak to peak rate for the UK, ie 2.25%. "Low fuel price" refers to the assumption that by 2000 fossil fuel prices will be slightly below their 1990 level in real terms.)

3.11 The current target being used in this programme is, therefore, a reduction in CO_2 emissions of around 10 MtC by 2000, to aim to return emissions to their 1990 level.

3.12 The EP 59 range (see paragraphs 2.17 to 2.19) and the target will need to be kept under review in the light of updated information, as part of the monitoring process (see paragraphs 3.81 to 3.84). In the short term, deviations from the working assumptions, including GDP growth and fuel prices, are inevitable, and would not necessarily imply that the measures being implemented needed to be changed unless they indicated a change in longer term trends.

MEASURES TO LIMIT EMISSIONS

Main principles

3.13 The UK's programme to limit emissions of CO_2, the main greenhouse gas, is based on a partnership approach. A wide-ranging consultation process formed the basis of this approach (see Chapter One). In order to achieve savings cost-effectively and to stimulate action throughout the economy, the Government will be working with a wide range of business, voluntary and other organisations. All have a valuable role to play in raising awareness of climate change, and providing advice on how to achieve energy savings.

3.14 The UK had already set in hand plans to reduce CO_2 emissions before the signature of the Convention in June 1992. In the Environment White Paper, "This Common Inheritance"[4], published in 1990 the Government set out a strategy to meet the then target of returning CO_2 emissions to 1990 levels by 2005. This target was conditional on other countries taking similar action. This strategy included energy efficiency measures, a review of renewable energy strategy, measures to limit transport emissions and recognition that in the longer term increases in the relative price of energy and fuel would be needed. It also undertook to monitor the results of this action to ensure that the target reductions were achieved and, if possible, improved on. The Government subsequently brought forward its target to return CO_2 emissions to 1990 levels by 2000, provided other countries took similar action.

3.15 The UK is seeking to achieve savings as cost-effectively as possible. The Government's priority is therefore to exploit the potential that exists to improve the efficiency of energy use, since this can bring economic and social benefits as well as reducing CO_2 emissions. Accordingly, the budget of the Department of the Environment's (DOE) Energy Efficiency Office (EEO) has been progressively increased, and for 1994 to 1995 will be over £100 million – a 150% increase over three years.

3.16 In choosing instruments to stimulate improvements in energy efficiency, the Government is adopting a presumption in favour of economic instruments rather than regulation (although regulation will continue to have an important role to play in certain circumstances). Many of the actions described in this programme are designed to remove market barriers to energy efficiency, for example by providing better information about investment opportunities, or by ensuring that the prices paid by energy users better reflect environmental costs.

3.17 The programme is based on a projection of emissions which has been adopted as the reference scenario (see paragraphs 3.9 to 3.12). The Government anticipates that the measures described in the programme should reduce CO_2 emissions in 2000 by some 10 MtC, which under the reference scenario is sufficient to meet the objective of aiming to return emissions to their 1990 level by 2000. The programme will be kept under review and adjusted as necessary to ensure that the UK fulfils its commitments under the Convention. Figure 3e provides a summary of the expected impact of the programme on emissions from different sectors. Some 1.5 MtC of the savings are expected to arise from supply side measures in electricity generation (further encouragement of renewables and combined heat and power). The rest of the savings arise from demand side measures.

Figure 3e   Summary of CO_2 savings

Sector*	Expected reduction in emissions by 2000, MtC
Energy consumption in the home	4
– introduction of VAT on domestic fuel use	
– new Energy Saving Trust	
– energy efficiency advice/information including Helping The Earth Begins At Home publicity	
– eco-labelling	
– EC SAVE programme (standards for household appliances)	
– revision of Building Regulations to strengthen energy efficiency requirements	
Energy consumption by business	2.5
– energy efficiency advice/information: Making A Corporate Commitment Campaign Best Practice Programme Regional Energy Efficiency Offices Energy Management Assistance Scheme	
– Energy Saving Trust schemes for small businesses	
– Energy Design Advice Scheme	
– possible EC SAVE programme (standards for office machinery)	
– revision of Building Regulations to strengthen energy efficiency requirements	
Energy consumption in the public sector	1
– targets for central and local government and public sector bodies	
Transport	2.5
– increases in road fuel duties and commitment to real increases of at least 5% on average in future Budgets	
Total	10

Note: * Savings in the electricity generating sector have been allocated to final users (including encouragement of renewables and CHP).

This table is intended as a summary of the key measures. Allowance has been made for overlap between some of the individual programmes.

Source: DOE

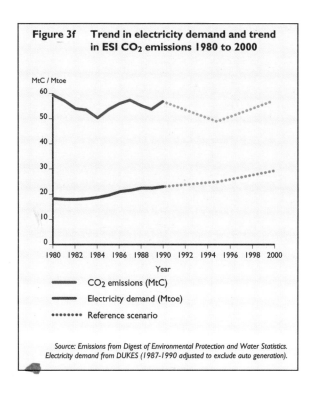

Figure 3f Trend in electricity demand and trend in ESI CO_2 emissions 1980 to 2000

— CO_2 emissions (MtC)
— Electricity demand (Mtoe)
······ Reference scenario

Source: Emissions from Digest of Environmental Protection and Water Statistics. Electricity demand from DUKES (1987-1990 adjusted to exclude auto generation).

Electricity Generation and Supply

3.18 The electricity generation and supply industry was responsible for about a third of UK CO_2 emissions in 1990. Under the reference scenario, despite a rise in electricity supplied, these emissions fall slightly both in absolute terms and as a proportion of the total by 2000 (see Figure 3f).

> **POWERGEN CO_2 TARGET**
>
> Of the two major fossil-fired generating companies in England and Wales one, **PowerGen**, is participating in the programme by setting a target aimed at ensuring CO_2 emissions from their electricity generating plant do not exceed 1990 levels by the year 2000. By switching from coal-fired to gas-fired CCGT plant and by constructing Combined Heat and Power (CHP) and renewables plant, PowerGen expect to contribute 3 MtC annually to the fall in CO_2 emissions. (These savings are already included in the reference scenario).

3.19 In recent years the UK electricity industry has undergone restructuring. Most of the industry is now in the private sector and a competitive regime has been established. This has increased the incentives for the industry to find the most efficient means of generating and supplying electricity. In particular, the

ELECTRICITY GENERATED BY NUCLEAR ENERGY

The generation of electricity by **Nuclear Electric** currently saves a substantial amount of CO_2 emissions. For example, if the electricity currently generated by nuclear power stations were to be generated by gas or coal, UK emissions would increase by some 6 to 15 MtC per annum. This is equivalent to between 4% to 10% of current CO_2 emissions. Recently Nuclear Electric have improved the performance of their Advanced Gas-cooled reactors and their expectations of future performance. This, together with the extra electricity supply from Sizewell "B", would increase the CO_2 benefits of nuclear electricity by further significant amounts. (These savings are already included in the reference scenario).

The electricity supply industry in Scotland makes a significant contribution to limiting CO_2 emissions. The high proportion of nuclear generated electricity in Scotland is a key factor. For example, if the electricity currently generated by **Scottish Nuclear Limited's** power stations were to be generated by gas or coal, UK emissions would increase by some 2 to 4 MtC a year. This is equivalent to 1% to 3% of current emissions. (These savings are already included in the reference scenario).

industry has been investing in high efficiency gas-fired combined cycle (CCGT) plant. CO_2 emissions per unit of electricity from this type of plant are about half those from conventional coal fired plant, due partly to greater efficiency and partly to the lower carbon content of gas. The reference scenario assumes that just under 12 gigawatts (GW) of CCGT capacity will be in place in the electricity supply industry by 2000.

3.20 The nuclear generating industry remains in public ownership. Around 20% of electricity generated in 1990 was from nuclear power, using two types of reactors: Magnox reactors built in the 1960s and Advanced Gas-cooled reactors built in the 1970s and 1980s. One pressurised water reactor is under construction, Sizewell "B". The Magnox stations are likely to be closed progressively over the next decade as they reach the end of their planned life. The construction times of nuclear reactors are such that no new stations could generate electricity before 2000, apart from Sizewell "B" which is scheduled to supply electricity from mid-1994.

3.21 In 1992 around 2% of UK electricity supply was generated from renewable energy sources, chiefly hydro power, with heat-producing renewables making additional contributions to the UK's overall energy requirements. Government policy is to stimulate the development of new and renewable energy sources wherever they have prospects of being economically attractive and environmentally acceptable in order to contribute to:
- diverse, secure and sustainable energy supplies;
- reduction in the emission of pollutants; and
- encouragement of internationally competitive industries.

3.22 The programme which implements this policy aims to stimulate the development of technologies and industrial and market infrastructure so that new and renewables are given the opportunity to compete equitably in a self-sustaining market. A major element of the programme is the Non-Fossil Fuel Obligation (NFFO) for renewables, and this is the main mechanism by which the Government will be working towards the figure of 1500 Megawatts (MW) of new renewable generating capacity in the UK by the year 2000. A third NFFO Renewables Order is proposed for the twelve Public Electricity Supply Licensees in England and Wales for approximately 300–400 MW of new capacity (this will be reviewed in the light of the quality and cost of proposals received). Separate bands are being considered for wind power, hydro, landfill gas from existing sites, municipal and industrial waste, energy crops and agricultural and forestry wastes. Crops for energy production are already encouraged by grants for their establishment and by special arrangements in the "set aside" provisions of the EC Common Agricultural Policy. A fourth Order is planned to come into effect in 1996 for about the same amount of capacity as the third Order. A further Order is planned to come into effect in 1998. The purpose of these Orders is to create an initial market so that in the not too distant future the most promising renewables can compete without financial support. This requires a steady convergence under successive Orders between the price paid under the NFFO and the market price. This will only be achieved if there is competition in the allocation of NFFO contracts.

3.23 In Scotland some 15% of electricity needs already come from renewable sources – predominantly hydro-electricity. Nevertheless, the Government intends that the two Scottish public electricity supply companies be required to obtain more electricity from renewable sources. The first Order under this proposed Scottish Renewables Obligation (SRO) is intended to be made in 1994. Further Orders are likely to be made in 1996 and 1998.

3.24 The other main element of the Government's strategy is the Department of Trade and Industry's (DTI) programme of research, development, demonstration and dissemination on new and renewable energy. This Government's support for research, development and demonstration into renewable sources has amounted to over £224 million since 1979. Resources are concentrated on key technologies such as biofuels, wind, solar and fuel cells, removing market barriers, and technology transfer. There is little scope in the UK for further large scale hydro power developments.

3.25 The increase in the figure which the Government is working towards for renewable electricity generating capacity is expected to save around 0.5 MtC towards the target of 10 MtC, spread across all sectors.

3.26 The Government is encouraging the use of CHP. This technology uses the waste heat from electricity generation to maximise the proportion of useful energy extracted from the primary energy source, the conversion efficiency. CHP plant can achieve conversion efficiencies of between 70% and 90% (compared with 35%–50% from conventional plant) provided a suitable heat load is available. In 1990 there was about 2000 MW of CHP capacity in the UK and the Government set a target of doubling this by the year 2000. At present there are approximately 1000 schemes in operation with a total installed capacity of over 2500 MW. Capacity currently being constructed should take this to 3000 MW by the

Figure 3g Electricity generated by type of plant in 1990 and 2000 under the reference scenario

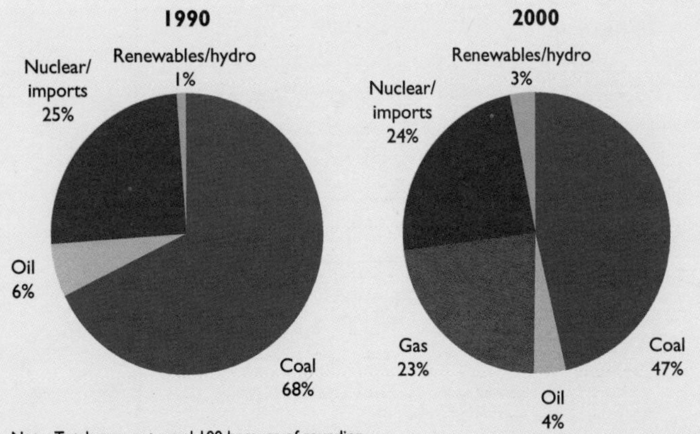

1990
- Nuclear/imports 25%
- Renewables/hydro 1%
- Oil 6%
- Coal 68%

2000
- Nuclear/imports 24%
- Renewables/hydro 3%
- Gas 23%
- Oil 4%
- Coal 47%

Note: Totals may not equal 100 because of rounding.

Source: DUKES and DTI Energy Paper 59

end of 1994. Given this successful growth the Government announced that it would increase this target to 5000 MW on the same timescale. This is expected to save about 1 MtC towards the target of 10 MtC, spread across all sectors.

3.27 The Government will be working with the CHP supply industry, CHP users, the CHPA (see box), and the Office of Electricity Regulation (OFFER) to remove market barriers and to promote the use of CHP technology wherever it is cost-effective. In particular, the EEO will use the Best Practice Programme (see paragraph 3.52) to promote CHP, providing independently monitored information on CHP systems to a targeted audience of potential users. OFFER and DTI keep the privatised electricity regime under review, and DTI have recently announced that those who generate their own electricity who consume less than half of their output on their site will generally not now require supply licences. This change (and others presently being considered by DTI) should further improve the prospects for on-site CHP generation. A database of all CHP schemes has been created by OFFER and is open to the public.

COMBINED HEAT AND POWER

The **Combined Heat and Power Association (CHPA)** is a professional Association working to promote the efficient use of energy through the wider use of CHP and community heating. With over 100 members, the Association has seen a steady increase in support and is fully committed to ensuring that the barriers to the wider use of CHP are addressed.

Since the Government created its initial CHP target of 4000 MW in 1990, the CHPA's members have contracted for 1000 MW of CHP capacity. In June 1993, a pilot programme was launched with the Energy Saving Trust (see paragraph 3.36) to secure the wider use of CHP in the residential sector. A budget of £1 million will fund at least 20 schemes in the first year. The CHPA is also working with the DOE to encourage local authorities to include CHP in their own housing energy efficiency strategies.

Energy Consumption In the Home

3.28 In 1990 energy consumption in the home accounted for just over a quarter of CO_2 emissions. (This figure includes CO_2 emissions attributable to electricity used in the home). Under the reference scenario these emissions were estimated to remain roughly constant in absolute terms, decreasing slightly as a proportion of the total (see Figure 3h).

3.29 The programme reflects the fact that CO_2 emissions from energy use in the home are the result of many separate decisions by households and individuals. Measures in the programme are aimed principally at helping and encouraging citizens to improve the efficiency of their energy use.

3.30 A key element of the programme is ensuring that appropriate price signals are given to households. Domestic fuel and light prices are on average 17% cheaper now than in 1983 in real terms, reflecting the fall in international energy prices and changes in the structure of the UK energy market, which have stimulated more competitive pricing. In March 1993 the Government announced that it would be introducing Value Added Tax (VAT) on domestic energy, at 8% from April 1994 and the standard rate (currently 17.5%) from April 1995. This will bring the UK into line with other European countries who already have a tax on domestic energy. Additional help will be available for those who are most vulnerable to the change. By encouraging more economical use of power and fuel, and further investment in energy efficient equipment in the home (which is already subject to VAT), the introduction of VAT is expected to result in a saving of about 1.5 MtC towards the 10 MtC target.

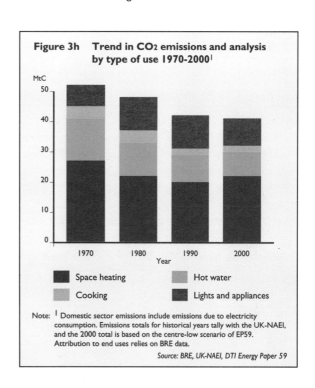

Figure 3h Trend in CO_2 emissions and analysis by type of use 1970-2000[1]

Note: [1] Domestic sector emissions include emissions due to electricity consumption. Emissions totals for historical years tally with the UK-NAEI, and the 2000 total is based on the centre-low scenario of EP59. Attribution to end uses relies on BRE data.

Source: BRE, UK-NAEI, DTI Energy Paper 59

3.31 The second main element, complementary to the first, is the provision of advice and information, both to increase awareness of the need to take action to reduce CO_2 emissions and to advise on the most cost-effective actions. In Autumn 1991 the EEO launched a new campaign, "Helping the Earth Begins at Home", aimed at raising the awareness of the link between the threat of global warming and energy use in the home, and providing information on the cost-effective measures that citizens and householders can take. The campaign has a total budget of £15 million over a three year period and has included press and television advertisements. Respondents to the advertisements receive authoritative advice on energy efficiency in the home, its cost, and the resulting savings in CO_2 emissions and energy bills. Further details of this campaign and other initiatives to increase public awareness of the threat of climate change are set out in Chapter Eight.

3.32 The Government is part funding with the Energy Saving Trust (see paragraph 3.36), a three-year pilot study of local energy advice centres providing free, independent and authoritative advice to the domestic and small business sectors. The aim of the study is to test the cost-effectiveness of providing local advice and information in increasing the take up of energy efficiency measures. Thirty centres were launched in October 1993. If the study proves successful the scheme could be extended nationwide.

3.33 In July 1993 the Government announced that it would increase the resources devoted to the provision of advice and encouragement to households. Resources for this purpose have been provided for in the recent public expenditure round.

3.34 Providing information on the energy consumption of individual household appliances and of houses and flats is important to enable consumers to choose the most energy efficient product for their circumstances. In September 1992 the European Community (EC) agreed a framework directive to require labelling of appliances. The UK hopes that the first implementing directives, covering refrigerators and freezers, will be in operation early in 1994. In the meantime, the public electricity supply companies have agreed a voluntary scheme to label refrigerators and freezers sold through their shops. A number of manufacturers have agreed to label their refrigerators and freezers sold in the UK.

PROVIDING ADVICE TO HOUSEHOLDS

Many other organisations are involved in the provision of advice and information to households. Both **British Gas and the regional electricity supply companies** have set up telephone advice lines on energy efficiency, including advice on the efficient use of fuel and appliances, under their Codes of Practice on energy efficiency.

The **Consumers' Association**, an independent non-government funded body providing advice and information to domestic consumers, is keen to promote information on CO_2 and energy reduction to consumers. In June 1993 it published energy advice on home energy surveys and carried out a survey on the quality of advice on energy efficiency available from major retailers in the UK.

ECO LABELLING

The Government has played a leading role in developing the EC ecolabelling scheme which will enable consumers to identify those products which are less harmful to the environment. To qualify for an ecolabel, products will need to meet stringent environmental criteria based on a life cycle analysis of the environmental impacts of that product group. The first product criteria, those for washing machines and dishwashers, were developed by the UK Ecolabelling Board and were approved in June 1993. The life cycle analysis of these products identified energy consumption during use as being one of the primary environmental impacts and the criteria therefore include limits on this. Other ranges currently being studied include refrigerators, batteries, light bulbs and insulation materials.

ACTION BY ENVIRONMENTAL GROUPS

Environmental groups are active and influential in the UK. Many provide information on the greenhouse effect and the environmental consequences of energy use to individuals. **Friends of the Earth** have produced a booklet, "Take the Heat Off the Planet", which provides a step-by-step guide to the action that individual householders can take to help reduce CO_2 emissions. **World Wide Fund for Nature** has produced a video and educational pack – aimed at 11–14 year olds – on the importance and nature of the Greenhouse Effect.

UTILITY REGULATION AND ENERGY EFFICIENCY

The UK has established independent regulatory authorities to oversee the gas and electricity utilities to protect competition and the interests of consumers. The Office of Gas Supply (OFGAS) and the Office of Electricity Regulation (OFFER) regulate the prices that the gas and electricity utilities can charge to domestic and small business customers and ensure that the utilities do not exploit their monopoly or near monopoly position.

Both OFGAS and OFFER have recently made changes to the way in which they control prices, in order to ensure that the utilities have an incentive to encourage energy efficiency where this is for the benefit of consumers. **OFGAS** has introduced an "E Factor" into the tariff formula which determines domestic gas prices. The "E Factor" allows expenditure by British Gas on approved energy efficiency schemes to be recovered by British Gas in the tariffs charged to gas consumers. This is already providing funds for schemes run by the Energy Saving Trust.

OFFER has announced a new supply price control formula for Regional Electricity Companies (RECs) in England and Wales to come into effect in April 1994. This will remove the previous incentive to maximise sales of electricity. OFFER will also set new Standards of Performance to require the RECs to explore the potential for cost-effective demand side management projects, and to promote the efficient use of electricity by undertaking projects in liaison with the Energy Saving Trust. A special revenue allowance will be made for such projects in setting the price control. This will finance nearly £100 million of new expenditure on energy efficiency over four years. An equivalent review of the supply price control formula for Scotland has just been put in hand.

Both OFGAS and OFFER were consulted on the establishment of the Energy Saving Trust and are contributing to its success.

3.35 A further EC directive was agreed in June 1993 requiring EC Member States to draw up and implement programmes to encourage energy efficiency improvements in a number of specified areas. The Government is already taking action to promote energy efficiency in all the areas required by the directive. In the domestic sector these include action on home energy labelling where two private sector organisations currently provide this service. A standard assessment procedure (SAP) has been introduced to enable the labels to be compared.

3.36 The third main element of the programme is the provision of financial incentives. In November 1992 a new Energy Saving Trust was established by the Government, British Gas and the public electricity supply companies. The Trust's aim is to promote the efficient use of energy, focusing initially on the domestic sector in particular. It has set itself a target of achieving savings of at least 2.5 MtC by 2000. Four schemes have already been launched. The first two are funded by British Gas via the "E Factor" (see box above): one provides grants towards the cost of high efficiency gas condensing boilers, and the other provides financial assistance towards the installation of CHP systems in residential properties. The third scheme is a pilot study of Local Energy Advice Centres (see paragraph 3.32). The fourth scheme, run in conjunction with the Lighting Industry Federation, provided financial incentives to the purchasers of low energy light bulbs. Detailed planning and consultations are currently underway to develop further schemes aimed at improving the energy efficiency of domestic dwellings. The energy saving measures that will be promoted will include insulation, heating controls and cavity wall filling. These schemes will be targeted to individual sectors of the market and launched in 1994.

3.37 Financial help is also targeted specifically at low income households through the Government Home Energy Efficiency Scheme. This provides advice and grants towards basic energy conservation measures. The budget for the scheme is to be almost doubled for three years from 1 April 1994, and the scheme extended to all pensioners and disabled people. The House Renovation Grant system, administered by local authorities, also provides help for energy efficiency to those on low incomes.

3.38 The fourth and final element of the programme is the setting of appropriate standards for new buildings and appliances. The Government has consulted on strengthening the building regulations as they apply to the domestic sector. The aim is to publish the revisions in early 1994 and introduce changes later in the year. It is estimated the proposals in domestic and non-domestic buildings would save 0.25 MtC.

3.39 The UK takes the view that where standards for traded goods such as domestic appliances are appropriate they should be applied within the EC as a whole. The objective of the first stage of the introduction of minimum standards is to give, by 1997, an average improvement in the efficiency of domestic electrical appliances of 10%, worth £100 million a year of total electricity savings to UK domestic consumers alone, by removing the most inefficient appliances from the market. A further toughening of standards to yield a 40% total improvement by the year 2000 is proposed. A directive on minimum efficiency standards for new hot water boilers was agreed in May 1992 and will be implemented by January 1994. The UK is pressing for other minimum standards to be agreed as soon as possible, with domestic refrigerators and freezers the priority, followed by washing machines, tumble driers and dishwashers. Savings from appliance standards could contribute 0.5 MtC towards the target.

3.40 Around 20% of UK housing is owned by local authorities. A further 3% is owned by housing associations, which are non-profit making organisations who receive public funds through the Housing Corporation, a Government agency which supports social housing in England. About one third of new homes are built by housing associations. Local authorities, housing associations and the Housing Corporation are all taking action to improve the energy efficiency of this housing as part of the UK's CO_2 programme.

3.41 The Government expects all local authorities to take fuller account in future of energy efficiency in their housing improvement strategies and programmes for their own housing, and to reflect this in their annual bids for expenditure, drawing on the lessons of the successful Green House Demonstration Programme. The Green House Programme promoted both proven energy efficiency measures combined in "packages" and more innovative measures including CHP. The 130 schemes involving packages of measures have the potential to save an average 55% per year of CO_2 (including an allowance for additional comfort). The Government has published interim guidance on good

ASSOCIATION FOR CONSERVATION OF ENERGY GUIDANCE NOTE

The **Association for Conservation of Energy**, an independent association of companies involved in the energy conservation business, has produced a "Best Practice" guidance note to utilise ideas from the successful DOE Green House Programme, which achieved 40 to 50% CO_2 savings.

GREEN HOUSE PROGRAMME

As a result of the programme, a rural authority, **Newark and Sherwood**, has developed a 20 year strategy to save CO_2 across its housing stock and has designated Boughton as a pilot energy village to demonstrate CO_2 saving refurbishment across all tenures.

Small scale CHP schemes for groups of estates or tower blocks such as the Green House schemes in **Birmingham** and **Leeds** have especial potential for replication in public sector housing. The 13 CHP schemes in the Programme are estimated to save an average of 38% of CO_2 from this single measure.

ENVIRONMENT CITY PROGRAMME

The Environment City Programme, launched in 1990, is an experiment to define what sustainability means in practice. The programme was devised by the Royal Society for Nature Conservation, The Wildlife Trusts Partnership, and the Leicester Ecology Trust, and is funded by British Telecom with support from the DOE. All four Environment Cities – in Leicester, Middlesbrough, Peterborough and Leeds – are pressing ahead on reducing CO_2 emissions through energy efficiency measures.

As the first authority to be designated as an Environment City in 1990, **Leicester City Council** is fully committed to reducing CO_2 emissions by 50% on 1990 levels by the year 2025 for its housing stock and local authority buildings, and through its transport responsibilities. The council is implementing a comprehensive Energy Action Plan to improve the energy efficiency of its own buildings and encourage the citizens of Leicester to reduce their energy consumption. The Eco-Feedback project encourages householders to record their energy use and set targets for energy consumption. Advice on how to meet the target is available from the Consumer Advice centre and by visiting Eco-House, Leicester's environment friendly show house. Leicester City Council has already carried out comprehensive energy surveys of its housing stock and aims to achieve a National Home Energy Rating of 10 (on a ten point scale) on all new housing.

ENERGY AUDIT BY WOODSPRING DISTRICT COUNCIL

Woodspring District Council's Housing Department, in Avon, has carried out an energy audit on all council houses and developed an energy strategy which will reduce CO_2 emissions from Council properties by a minimum of 10% per annum from March 1993 to March 1999.

GUIDANCE FROM FRIENDS OF THE EARTH

Friends of the Earth is producing guidance for local authorities on measures that they can adopt to reduce CO_2 emissions from their districts and make recommendations for further action by Government and others.

practice, including guidance on CHP schemes, drawing on the lessons from the Green House Programme. Further consolidated guidance will be published in 1994.

3.42 Local authorities are being encouraged to calculate energy ratings and use these to set targets for the energy efficiency of their housing stock. The Environment City programme is leading the way (see box on facing page).

3.43 The Housing Corporation and housing associations are now recommending an energy efficiency standard in new housing of at least SAP 75 on a 100 point scale (see paragraph 3.35).

3.44 The Corporation is also examining the scope for efficiency improvements in existing housing association stock. Early indications are that when housing associations are renewing heating systems, a simple SAP target of 60 should be achievable. However, energy costs (and hence CO_2 emissions) vary significantly with property size.

3.45 The Corporation is encouraging housing associations to calculate energy ratings for their properties as a first step to setting energy targets, and to work with the Energy Saving Trust where schemes are applicable to their housing.

3.46 Overall, after allowing for some overlap between the effects of different measures, and taking into account the impact of changes in the electricity generation and supply industry on CO_2 emissions attributable to electricity use, these measures to limit emissions from energy consumption in the home are expected to lead to savings of the order of 4 MtC towards the 10 MtC target.

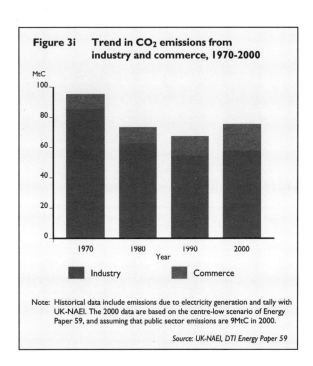

Figure 3i Trend in CO_2 emissions from industry and commerce, 1970-2000

Note: Historical data include emissions due to electricity generation and tally with UK-NAEI. The 2000 data are based on the centre-low scenario of Energy Paper 59, and assuming that public sector emissions are 9MtC in 2000.

Source: UK-NAEI, DTI Energy Paper 59

Energy Consumption by Business

3.47 In 1990 energy consumption by industry and commerce, excluding transport, accounted for some 43% of UK CO_2 emissions. Under the reference scenario, these emissions rise by 2000, with relatively strong growth in the commercial sector (see Figure 3i).

3.48 As with the domestic sector, emission totals are the result of a large number of individual decisions. However because of the need to maintain competitiveness in business the incentive to reduce costs through energy efficiency tends to be stronger than in the domestic sector, particularly in the more energy intensive industries. The programme aims to reinforce these inbuilt market incentives by encouraging business to take voluntary action to save energy.

3.49 There are two main elements to the programme. The first, the Making a Corporate Commitment Campaign (MACC) launched in October 1991, seeks top management commitment to energy efficiency from private and public sector organisations. The campaign asks organisations to give a board level director responsibility for developing and reassessing their energy efficiency strategy, to set performance improvement targets, and to ensure that plans are considered regularly at board level. Over 1600 organisations have already joined the campaign, and over 40 trade associations, including the Confederation of British Industry (CBI) and the Institute of Directors (IOD), have endorsed it and are promoting its message among their members.

3.50 While the first aim of the campaign is to stimulate action by organisations, its second aim is to promote the advisory services of the EEO – Best Practice Programme and Regional Energy Efficiency Officers (see paragraph 3.54). The campaign advocates a management approach to energy, and successful case-studies have been captured in a video 'Successful Energy Management' and in a booklet 'Energy, Environment and Profits', both of which have been disseminated widely. The campaign will be producing further guidance on fuel efficiency.

3.51 The UK was the first country to develop a standard for environmental management systems, BS 7750, which is due to be published in final form early in 1994 following extensive pilot testing. It shares common principles with BS 5750 (and its European and international equivalents; EN 29000 and ISO 9000) for quality systems. Energy management procedures, including target setting and progress monitoring, will in most cases be an important part of an environmental management system, and help companies to reduce energy requirements. BS 7750 is compatible with the European Eco-Management and Audit (EMA) scheme, adopted in June 1993. When fully operational in 1995 the EMA scheme will provide a European register of industrial sites committed to continuous environmental improvement, with regularly audited environmental management systems in place, which publish independently verified environmental statements. Proposals for implementing the EMA Regulation in the UK have been published by the Government.

MAKING A CORPORATE COMMITMENT CAMPAIGN

The setting of performance improvement targets is a key part of the MACC campaign. In September 1993 the Secretary of State for the Environment wrote to all organisations that have joined the campaign asking them to publicise their energy efficiency targets. In November the Secretary of State hosted a lunch for key signatories inviting them to play a leading role in the CO_2 partnership. A number of companies have declared the targets they intend to meet. These include:

3M: A global target of a 20% reduction in energy use to produce a kilogram of good product from 1990 to 1994; and a 20% reduction in the amount of energy required to support a square metre of non-manufacturing space from 1990 to 1995;

Allied Lyons plc: Following significant earlier savings, the company has set energy, effluent and water savings targets of around 2% per annum, over a three to five year timescale;

BAA plc: A 10% reduction in consumption in all airport operations by December 1994 with the expectation of a further 5% saving in the following four years; and an initial 10% improvement in non-airport operations by January 1995;

BT plc: A 15% reduction in energy consumption over five years to 1997;

Babcock Energy Ltd: A 10% reduction in energy costs in 1993/94;

Bank of England: A 15% reduction in energy consumption over five years to 1998;

Birmingham International Airport plc: As one of its energy targets, a further 5% reduction in Kwh per passenger, based on the moving annual total;

Bowring Services Ltd: A minimum 5% annual reduction in energy costs;

British Gas plc: A 15% saving over five years;

Cheltenham & Gloucester Building Society: Twin goals to reduce energy costs by 10% within two years and energy consumption by 10% by the end of 1996;

DHL International (UK): An improvement in fuel efficiency of 15% over three years commencing with a target of 8% in the year October 1993/94, and a 15% improvement in energy use at operational sites by the end of 1994, rising to 20% in 1995 and 25% in 1996;

Eley Ltd: A 5% reduction in energy consumption in 1993, taking production volume changes into account;

English Nature: A 15% reduction in energy consumption over five years;

Fina plc: A performance improvement target for energy costs of 10% and a 5% reduction in energy consumption for individual areas of the company over a 2-year period ending December 1994;

Forte: An initial target to reduce energy and water consumption by 10% without reducing the level of service, quality or customer comfort;

IBM: A worldwide corporate goal of implementing energy efficient improvement projects to achieve 4% savings year on year;

J. Sainsbury plc: A target of a 5% annual saving on the difference between actual energy used and the predicted energy which should have been used if utilised in the most efficient manner;

Nationwide Building Society: A 15% target over five years with the expectation that this could be increased to 20 to 25% in the longer term;

National Westminster Bank plc: A 15% reduction in energy consumption by the end of 1995;

Rockware Glass Ltd: CO_2 emissions in 2000 will not exceed 1992 values;

Rover Group: A 5% per annum reduction in 1991 energy and CO_2 levels adjusted for the number of standard production hours worked and degree days during the year;

Safeway Stores plc: A real saving of at least 10% over five years;

Sterling Organics: A programme of improvement projects which will reduce energy consumption by 6% in the two years ending 1994, against a baseline of 1992;

Tesco plc: A further 5% reduction in energy use in each of the next two years;

Texaco Ltd: Improve the energy index at its Pembroke refinery by 1% each year over the next five years; and save 15% of its energy consumption in its H.Q. building in 1994;

Toshiba (Manufacturing Operations): A 16% improvement in energy use by 1995;

Warwick International Ltd: A 20% improvement in energy use per unit of product produced in 1993 and 1994;

University College London: A 15% reduction in effective energy use over five years to 1998;

ELECTRICITY COMPANIES

Manweb plc: A 10% reduction in own electricity use in the three years to 1994/1995;

Norweb plc: A 5% reduction in own electricity consumption over 5 years to 1996;

Northern Electric plc: To reduce own energy usage by one-third over the next five years;

Nuclear Electric plc: A 5% reduction in overall energy bill in 1993;

Southern Electric plc: A 20% reduction in energy consumption in office buildings over next five years;

LOCAL AUTHORITIES

Basingstoke & Dean Borough Council: 15% reduction in effective energy use in five years to 1998;

Bedfordshire County Council: 15% reduction in energy costs over five years to 1991 and then 10% by 2000;

Cleveland County Council: 7.5% for educational establishments while Cleveland Police Authority have set a target of 10% over one year for their buildings;

Cotswold District Council: 15% over five years to 1997;

Ipswich Borough Council: 15% over five years to 1998;

Leicester City Council: To complement a 14% reduction in CO_2 emissions over the past five years, the Council intends to reduce CO_2 emissions by 50% of the 1990 level by the year 2025;

London Borough of Hounslow: To reduce 1992 energy consumption by 20% by the end of 1997, with consequent CO_2 emission reductions in excess of 20%, by investing 10% of the annual energy budget in a programme of building and plant improvements, and by encouraging and implementing responsible energy management measures;

London Fire and Civil Defence Authority: A 10% reduction in energy consumption over five years to 1994/95;

London Borough of Sutton: 4% energy savings in this financial year and a 15% saving over 10 years;

Northamptonshire County Council: To reduce energy and water consumption and CO_2 emissions by a further 15% over five years commencing April 1994;

Nottingham City Council: A 10% reduction in energy consumption by 1994/95; a further 5% by 1996/97 with up to a 50% reduction by the year 2027;

Oxfordshire County Council: On course to reduce their energy use in Council buildings by 15% over the five years to 1998;

Redditch Borough Council: A 15% reduction in energy costs over three years to 1995;

Stratford on Avon District Council: A 5% reduction in the 1992/93 level of normalised energy consumption by March 1995;

Suffolk Coastal District Council: Following cumulative reductions to date of some 5,000 tonnes of CO_2 emissions, the Council intends to reduce energy consumption by a further 10% in the five years to 1998; and

West Sussex County Council: A CO_2 reduction target of 50,000 tonnes per year by 2000.

3.52 The second main element of the programme is the provision of information and advice to companies on how to improve energy efficiency. The EEO runs the very successful Best Practice Programme. This has four elements:
- Energy Consumption Guides which give data to enable users to establish their relative energy performance within their sector;
- Good Practice Guides and Case Studies, which give independent information on proven energy saving measures and techniques and what they are achieving;
- New Practice Projects which give independently monitored performance data on new energy efficiency measures which do not yet enjoy a wide market; and
- Future Practice which supports research and development into new energy efficiency good practice measures.

The information and publications from the programme are disseminated by direct mailings to contact points in business and the public sector and through workshops, seminars and site visits.

3.53 The small business sector uses nearly half the total energy used by industry and commerce. Unlike larger organisations, however, many small companies do not have a management structure which can appraise energy management issues. In April 1992 the Government launched a new Energy Management Assistance Scheme (EMAS) to provide financial help to smaller companies with less than 500 employees to enable them to obtain consultancy advice on the design and implementation of energy efficiency projects. In 1993–94 EMAS is expected to pay grants totalling £1.7 million to over 3,000 small businesses in the UK. It is anticipated that this will promote annual energy efficiency savings of 5 times the expenditure on grants. In addition, a new Energy Saving Guide for small businesses has been published in conjunction with the Advisory Committee on Business and the Environment (ACBE). Energy Saving Trust programmes for this sector will also be developed.

3.54 Energy efficiency advice is also available to companies direct from eleven Regional Energy Efficiency Officers (REEOs) based in offices around the country. REEOs and

---

BUSINESS GROUPS ARE ENCOURAGING TARGET SETTING

Business groups and trade associations are also encouraging companies to set demanding energy saving targets. Members of the **Chemical Industries Association** (CIA) have signed up to their trade association's "Responsible Care" programme, which has been running for four years in the UK as part of a worldwide programme of voluntary regulation, involving top management commitment and the setting of management standards. Within that programme, CIA launched in March 1992 a "Responsible Energy" initiative (a joint exercise with the EEO) to assist its member companies to make further improvements in energy efficiency. In June 1993, CIA led the way among businesses by publishing performance indicators, including an indicator of energy efficiency, to demonstrate achievement.

The **British Ceramic Confederation** (BCC) has adopted an environmental strategy and issued guidelines to all its members in March 1993. As part of this strategy, BCC is monitoring the energy efficiency of the ceramics industry and is considering encouraging its members to set specific energy consumption targets.

The **British Combustion Equipment Manufacturers Association** (BCEMA) is keen to help reduce CO_2 emissions from the industry. BCEMA has endorsed the MACC campaign and is asking all of its members to make a formal commitment to energy efficiency targets. In addition, BCEMA has circulated energy policy guidance to its members and customers with the aim of reducing energy consumption to a level 10% below 1990 levels.

The Government has set a target of a 50% recycling rate for all recyclable household waste by the year 2000. **British Glass**, the trade body for the UK glass industry, is committed to meeting this target for all glass containers and is running an active PR and advertising campaign to encourage more glass recycling by industry, local authorities and government. Producing glass by recycling can reduce CO_2 emissions.

their staff collectively visit a total of around 2500 energy users, and participate in over 300 exhibitions, conferences and seminars each year. A series of high level seminars and exhibitions is planned for the Spring of 1994, to inform organisations about the CO_2 programme, encourage target setting and provide a forum for organisations to obtain advice and information, to enable them to take action.

DISSEMINATION OF ADVICE BY BUSINESS GROUPS

Business groups and trade associations also play an important role in the dissemination of advice to companies. Through the Confederation of British Industry (**CBI**) Environment Business Forum, companies are being encouraged to take positive steps on environmental issues. Members of the Forum are committed to public reporting on environmental performance as a means to establishing and maintaining stakeholder confidence in their operations. Recent initiatives include a series of seminars based around CBI's energy management guidelines in addition to the ongoing dialogue with member companies and trade associations on the scope for further business-led voluntary action.

ENCOURAGEMENT TO TAKE ACTION BY BUSINESS GROUPS

The Institute of Directors (**IOD**), the **British Chambers of Commerce** and the **British Retail Consortium** are encouraging member companies and trade associations to take action on environmental issues. The IOD published a guide on energy efficiency in September 1991 and surveys on environmental motivations are regularly published.

3.55 The Government is also supporting ten local 'green business' initiatives, in partnership with the private sector, on the recommendation of ACBE. These initiatives will raise the awareness of energy and other environmental issues among small and medium sized companies and offer them practical help and guidance on improving their performance. Nine are already operational – in Amber Valley, Blackburn, Dudley, Hertfordshire, Leeds, Newcastle, Plymouth, Sheffield, and Wearside.

3.56 Resources provided for in the recent public expenditure round will allow for further strengthening of these programmes of information and advice. The targets for the Best Practice Programme will be further revised in due course in the light of detailed surveys of the impact of the programmes.

ENERGY SAVING INITIATIVES IN WALES

In Wales, the Energy Managers' Groups meet around three times a year and provide an active forum for disseminating information and advice on the elimination of energy waste and the better planning of energy use. This includes helping to identify opportunities for investment in energy saving techniques.

For the past year, an initiative supported by the Welsh Office has been directed at industry to stimulate awareness of the constraints and opportunities presented by environmental pressures. It aims to promote best environmental practices and provide practical advice to industry. The initiative also works closely with the EEO.

With the aid of a grant from the EC, the Welsh Development Agency set up the Welsh Energy Project in 1991, to study the energy flows and use in Wales. The final report was completed in April 1993 and will assist organisations, including the EEO (Wales), in promoting energy efficiency.

3.57 As with the domestic sector, energy efficiency standards have a part to play. The Government has consulted on strengthening the building regulations as they apply to the commercial/industrial sector. The aim is to publish the revisions in early 1994 and introduce changes later in the year. It is estimated the proposals in domestic and non-domestic buildings would save 0.25 MtC. The directive setting minimum efficiency standards for new hot water boilers also covers commercial sizes up to 400 kW from January 1994. The European Commission is also undertaking studies under SAVE (Specific Actions for Vigorous Energy Efficiency) of the possible benefits of minimum efficiency standards for office and other commercial equipment.

3.58 In 1992 the Government launched the Energy Design Advice Scheme (EDAS) to help ensure that the design of building projects, both new build and refurbishment, is undertaken with full awareness of the latest proven energy saving techniques available. This scheme will help achieve further CO_2 reductions.

3.59 Overall, taking account of changes in the electricity generation and supply industry, the programme is estimated

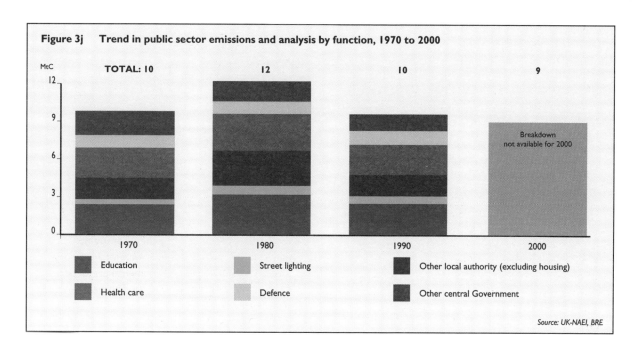

Figure 3j Trend in public sector emissions and analysis by function, 1970 to 2000

to lead to savings of some 2.5 MtC in emissions attributable to energy use by industry and commerce, towards the 10 MtC target.

Energy Consumption in the Public Sector

3.60 In 1990 energy use by the public sector was responsible for about 6% of UK CO_2 emissions. These emissions were expected to fall by 2000 under the reference scenario (see Figure 3J).

3.61 The UK Government takes the view that the public sector should provide a lead to industry and commerce by setting itself demanding energy saving targets and publicly reporting progress. In 1990 the Government set itself a 15% energy saving target over the five years to 1995 to 1996.

3.62 Health authorities are also working towards a 15% target over a similar period and the local authority associations have endorsed a similar target for local government energy consumption in their non-housing buildings.

3.63 Further targets will be set in due course to cover the period up to the year 2000. The aim is to reduce energy use in the public sector by at least 20% compared with 1990 levels. This should save around 1 MtC towards the 10 MtC target.

ENERGY SAVING ACTION BY GOVERNMENT DEPARTMENTS

The performance of Government departments towards meeting the target of 15% energy savings over the five years ending 1995–96 is measured using three indicators. The first indicator "$£/m^2$" measures energy consumption at fixed prices, weather-corrected and adjusted for estate size. The results for 1992–93 show the non-defence estate, whose energy bill amounts to about £112 million a year, 0.2% better in performance terms than in 1990–91. Comprehensive data for the defence estate are not yet available.

The two additional indicators are "CO_2/m^2" and "total CO_2 emissions". On these, there has been a 6% improvement on a CO_2/m^2 basis, but a worsening of 2% in total CO_2 emissions. All figures exclude the defence estate.

Government departments have been encouraged to follow the recommendations of good practice in the use of their transport as set out in the guide published by the DOE in conjunction with ACBE.

3.64 Measures are being taken to ensure that the necessary investment to achieve these targets can take place. In central government, changes have been made to facilitate greater use of Contract Energy Management by departments and other government bodies. For local government, a programme of Supplementary Credit Approvals to the amount of £3 million has been made available in 1993–94 to allow capital investment in energy efficiency measures in authorities' general administrative buildings. The EEO is also working with local authorities through the Central and Local Government Environment Forum (CLGEF) and its working group on energy efficiency.

Transport

3.65 In 1990 transport accounted for around 24% of UK CO_2 emissions. This rises to 26% of emissions in 2000 under the reference scenario. Over half of the 1990 emissions came from private use of cars. Nearly 30% came from industrial and commercial road transport (see Figure 3k).

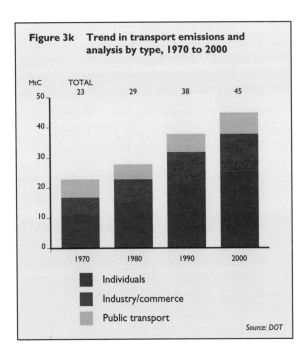

3.66 The programme aims to limit transport emissions by improving the fuel efficiency of vehicles and of driving patterns. As in other sectors a market based approach is being used, and a key element of the programme is providing the right price incentives. Transport fuel duties have been increased in real terms. In March 1993 the Government announced that it would continue this strategy in the longer term as part of the CO_2 Programme. Fuel duties were increased by 3 pence per litre (including VAT) in the November 1993 Budget, representing an increase of between 8% and 10% in cash terms. In future Budgets fuel duties will be increased by at least 5% a year on average above the rate of inflation (see Figure 3l). This measure is expected to save around 2.5 MtC towards the 10 MtC target.

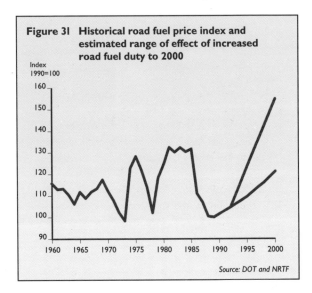

3.67 The Government has discussed with UK motor manufacturers and importers the scope for improving the fuel economy of vehicles sold in the UK in the light of the changing market conditions produced by the rise in fuel duties. A forum will be established between Government, motor manufacturers and other interested parties to co-ordinate marketing activities on "greener motoring".

3.68 There are a number of other initiatives in the transport sector which, although primarily designed to achieve other ends (such as reducing traffic congestion or improving road safety), will have a beneficial effect on CO_2 emissions. The Government is pursuing a range of measures to reduce the environmental impacts of transport and influence the rate of traffic growth. Measures are also being taken to encourage freight traffic to use rail rather than road transport and passengers to use public rather than private transport.

3.69 The Government is also examining the option of charging for the use of roads by vehicles. Research is under way into the possibility of pricing roads in London and other towns, and the environmental impact of different options. Research into city congestion charging in London is due to be completed by the end of 1994. In November 1993 the Government announced its intention to introduce electronic tolling on the motorway network when the technology is available. Motorway charging will help to deal with congestion from rising demands for road capacity as the economy grows.

IMPROVEMENTS IN VEHICLE EFFICIENCY BY CAR MANUFACTURERS

UK manufacturers are committed to meeting the Association of European Car Manufacturers (ACEA) target of a 10% improvement in vehicle efficiency by 2005.

Vauxhall has introduced a range of fuel economy models which are up to 12% more economical than the comparative standard versions. Vauxhall is also committed to a number of further initiatives, including an increase in its promotional activities related to fuel economy.

Since 1980 **Rolls Royce Motor Cars** has achieved a 15% reduction in the fuel consumption of their vehicles sold in the UK. Further improvements are being made and the company has set a technical strategy target of achieving a cumulative reduction of 25% in CO_2 emissions from its UK model range over the period 1990 to 2000.

Ford has introduced a product cycle plan, which includes fuel economy tracking, to produce a product range which will meet the ACEA commitment. In addition, Ford is currently achieving a 4% year on year reduction in CO_2 emissions from its manufacturing operations.

Rover has, through comprehensive energy management measures, reduced energy consumption in its vehicle manufacturing activities by nearly 10% in 1992, and is committed to a 25% reduction on 1991 levels by 1996. Rover products are also benefitting from new engine technologies which reduce fuel consumption and CO_2 emissions as part of the company's "whole life" environmental policy. The company is committed to further CO_2 improvements through technology and to raising the level of customer awareness of environmental issues.

It will improve the competitive position of public transport. Charges set too high would cause traffic to divert from motorways to other more sensitive routes, although it must be borne in mind that without action congestion from rising traffic levels on motorways would also cause diversion.

3.70 Bus and rail services can offer advantages where they are well loaded and provide an alternative to private car use. The Government is providing subsidies to maintain regional railway services and public transport in London. Local authorities subsidise some bus and rail services. The Government is also providing grants towards investment in facilities for transporting freight by rail. And all local bus services receive partial rebate for the duty paid on fuel. Public transport is zero-rated for VAT.

3.71 Lower speeds and a calmer driving style can also reduce emissions. Fuel consumption and hence CO_2 emissions increase as speed increases beyond 50 miles per hour (mph). They are also increased by hard acceleration. In November 1992 the Government issued a policy document setting out its strategy for tackling the problem of excess speed on the roads. This was primarily directed at reducing road casualties but the measures, which included more widespread use of camera technology to enforce speed limits, a campaign to change public attitudes towards speed and the fitting of speed limiters to most coaches and heavy goods vehicles, will also have benefits for the environment.

3.72 The Government is advising local planning and transport authorities to look at ways of reducing the need to travel and at alternatives to private car travel. New planning policy guidance is in preparation (see Chapter Ten). And transport authorities now have greater freedom to switch resources between different forms of transport.

TRANSPORT MEASURES BY LOCAL AUTHORITIES

York City Council have undertaken a co-ordinated package of measures designed to secure environmental improvements through promoting environmentally sound modes of transport while managing traffic pressures and restraining parking supply. The measures include the creation of one of Europe's largest pedestrian zones, the provision of park-and-ride services and a system of safe cycle routes.

Birmingham City Council is working with businesses and voluntary groups to raise awareness of the impact of transport on the environment in the city. A week of activities about transport is planned, culminating in a conference in March 1994.

Peterborough, the most recently designated Environment City, has introduced a cycling strategy which encourages people to leave their cars at home. Between 10 to 12% of the city's commuters regularly use their bikes to get to work.

ADVICE FROM ENVIRONMENTAL, BUSINESS AND CONSUMER GROUPS ABOUT TRANSPORT USE

Environmental and consumer groups have an important role to play in spreading awareness of the links between transport use and environmental damage. **Transport 2000** and the **Environmental Transport Association** jointly co-ordinated the first Green Transport Week in Great Britain in June 1993. A second Green Transport Week is planned for June 1994. The aim of the week is to raise awareness of the environmental impacts of transport, including its contribution to climate change, and to encourage people to take action to reduce these impacts. During 1994, Transport 2000 is also planning a series of courses for local councillors on implementing sustainable transport policies.

The **Automobile Association** and **Royal Automobile Club** are assisting in educating drivers' attitudes to fuel efficiency. In addition to articles in its members' magazine, the AA recently placed an advertisement on fuel efficient driving in a national newspaper and has published a list of 10 measures that motorists can carry out to drive more fuel efficiently. The RAC runs an environmental telephone advice service and publishes a leaflet giving advice on how to save fuel to both its members and the motoring public in general. The RAC has also established an independent foundation to promote research aimed at resolving the apparent conflicts between car use and environmental concerns.

Companies can help to reduce emissions by promoting good practice among their employees. **DHL International (UK) Ltd** is a division of a leading international distribution company. As part of its corporate environment policy, it is committed to reducing environmental impacts of transport and has set itself a target to improve fuel efficiency by 15% over three years, commencing with a target of 8% in the year October 1993–94. The target will be achieved through a combination of education, purchasing of more fuel efficient vehicles, reporting, monitoring and driver training. To drive this initiative forward, the company has produced for all staff a Fuel Efficient Driving Booklet which explains how to improve fuel efficiency and reduce other environmental impacts of driving a vehicle. To ensure that all staff participate in the fuel efficient driving scheme, DHL has established a performance league table which records improvements in fuel efficiency by departments and regional depots.

British Airways recently ran a campaign designed to raise awareness among the airline's Heathrow-based staff of the environmental benefits of car sharing to and from work. The driver of the car-sharing vehicle was given an energy-saving light-bulb and each occupant entered a prize draw.

The **Cyclist's Touring Club**, Britain's national cyclists' association, promotes the use of the bicycle for leisure and other purposes, as well as for improved public health and the environmental benefits, including the reduction of CO_2 emissions. As part of its promotional activities, CTC will be helping to co-ordinate a "National Bike Week" on 11–19 June 1994.

British Gas is promoting the use of compressed natural gas in vehicles because CO_2 emissions from such vehicles are up to 30% lower than those from petrol-driven vehicles. The company is converting 300 vehicles to run on compressed natural gas.

The **Government Car Service** and the **Government's Inter-Despatch Service** are carrying out trials on natural gas vehicles. If the trials are successful, and it is practical to do so, it is intended that the entire fleet of 260 vehicles will be converted to run on natural gas.

The **Freight Transport Association,** representing about 12,000 member companies, is assisting industry to devise a framework of achievable environmental targets. FTA has produced, in association with the EEO, a comprehensive guide to good fuel management and a video on improving vehicle aerodynamics. Proven techniques can give fuel savings: improved aerodynamics up to 20%; driver training, 6%; vehicle maintenance, 8%. FTA has also published the first Directory of Environmental and Freight Transport Research, giving details of over 160 current research projects on many subjects, including global warming. FTA is committed to ensuring that the transport industry continues to recognise its environmental responsibilities without jeopardising operational efficiency.

Esso has sponsored a reprint of the Department of Transport's leaflet "Motoring and the Environment", which encourages motorists to use less fuel and thus reduce pollution. The leaflet is being distributed within the company and to motorists throughout the network of Esso service stations.

3.73 UK local authorities are managing the flow of traffic in towns in a variety of ways. Measures include parking controls, pedestrian only zones, park-and-ride schemes where commuters can leave their car outside a town and catch a bus service into the centre, traffic calming to slow speeds, provision for pedestrians and cyclists and public transport priority measures.

3.74 The UK is making information available to consumers through publicity campaigns and publications about the importance of fuel economy. Consumers are encouraged to think carefully about their use of cars in the light of environmental impacts of road traffic.

3.75 In addition to the measures described in this section, the Commission of the European Community has been asked to put forward proposals for the reduction of CO_2 emissions from passenger cars. The Government hopes that constructive proposals will come forward in the near future.

COSTS AND ECONOMIC IMPACT

3.76 Most of the measures in the programme are designed to improve the energy efficiency of the UK economy, and also help to reduce costs. The use of economic instruments and of information and encouragement, in preference to regulation, will help to ensure that savings are made where they are most cost effective. Overall it is expected that the programme will improve efficiency and could lead to significant resource cost savings.

3.77 The promotion of renewable energy resources is part of a longer term strategy to assess renewable energy technologies and where appropriate develop them to the point where they can become economic and contribute to ensuring a secure and diverse energy supply in the UK.

Effect of programme on emissions trends 1990 to 2000

3.78 The path that emissions will take to meet the 2000 target cannot be estimated with precision. Fluctuations in economic activity and temperature patterns, among other things, can strongly influence the level of emissions in any one year and disguise both the underlying trend and the impact of the programme. Nonetheless we can illustrate how the programme might affect the underlying trend in emissions in order to provide a basis for monitoring progress.

3.79 Figure 3m shows the reference scenario which is used to represent the possible range for the longer term development of underlying emissions up to the year 2000.

(This has been adjusted to take account of the changes in the 1990 estimate noted in paragraph 3.5). The lower line in the graph provides an approximation of how the longer-term trends represented by the reference scenario could be expected to change as a result of the programme.

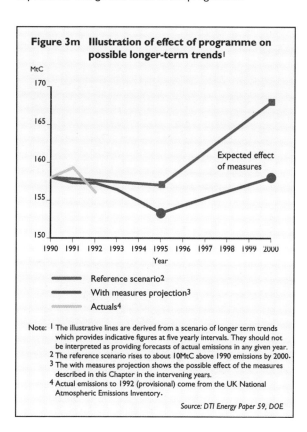

Figure 3m Illustration of effect of programme on possible longer-term trends[1]

Reference scenario[2]
With measures projection[3]
Actuals[4]

Note: [1] The illustrative lines are derived from a scenario of longer term trends which provides indicative figures at five yearly intervals. They should not be interpreted as providing forecasts of actual emissions in any given year.
[2] The reference scenario rises to about 10MtC above 1990 emissions by 2000.
[3] The with measures projection shows the possible effect of the measures described in this Chapter in the intervening years.
[4] Actual emissions to 1992 (provisional) come from the UK National Atmospheric Emissions Inventory.

Source: DTI Energy Paper 59, DOE

3.80 These illustrative lines can only provide a general guide to progress. In monitoring progress, year-on-year developments will be assessed in the context of the continuing analysis which underlies these illustrative lines. As can be seen from the actual emissions figures for 1991 and 1992 (provisional), emissions levels in the UK can vary significantly from year to year demonstrating the wide variation around the trend that can occur on an annual basis (see paragraph 3.82). This is partly due to fluctuations in circumstances such as the level of economic activity and temperature. Also, changes in the mix of fuels used will have an impact on the path of emissions.

MONITORING

3.81 The UK will monitor progress towards fulfilment of its commitments on CO_2 emissions through:
- annual updating of the emissions inventory and publication of emission estimates;

MONITORING IN 1990, 1991 AND 1992

UK CO_2 emissions in 1991 were estimated to be 159 MtC, about 1 MtC above the 1990 level. The provisional estimate for 1992 is 156 MtC, some 2 MtC less than emissions in 1990. Indications so far are that emissions in 1993 will be below the level in 1992.

Since 1990, reductions in economic output (GDP) and changes in the composition of fuels used have tended to reduce UK CO_2 emissions, whilst temperature variations have worked in the opposite direction (because 1990 was an exceptionally warm year). Economic activity in the United Kingdom declined by 2.5 per cent in 1991, compared to a year earlier, and by a further 0.5 per cent in 1992. (This compares with an average historical growth rate of about 2.25 per cent.) Economic growth has resumed in 1993. It is difficult to measure the precise impact of reduced economic activity on CO_2 emissions but it will certainly lead to a lower level of CO_2 emissions than would otherwise have been the case.

There has also, over the period 1990 to 1993, been an increase in gas consumption by the electricity generating industry and in generation of nuclear electricity which will both have contributed to lower levels of CO_2. On the other hand, average temperatures in the UK have been nearer to normal in 1991, 1992 and 1993 than in 1990, when the first quarter was particularly mild. This factor will have contributed to an increase in CO_2 emissions compared to 1990. As the programme gets underway the impact of measures in the programme can be expected to reduce emissions below the trends indicated by the reference scenario.

- regular evaluation of historical trends in emissions and the effectiveness of measures; and
- periodic reappraisal of longer-term future trends in emissions and the potential impact of policy measures.

3.82 Statistical evaluation of year-on-year changes in emissions helps explain short-term trends by separating out the estimated effects of temperature variations and fluctuations in GDP. It is also helpful to note major changes, such as in the composition of fuels used in electricity generation, for comparison against expectations. For assessing the longer-term outlook, more detailed analysis of trends over time, and consideration of longer-run constraints is necessary.

3.83 Annual data will continue to be published and used to monitor emissions trends in individual sectors and may provide insights into the impact of measures on particular sectors. For example, since 1990 CO_2 emissions from the industrial sector have declined, as the UK has gone through a period of recession. At the same time growth in CO_2 emissions from the transport sector has halted, as fuel consumption in that sector has remained broadly flat. CO_2 emissions from domestic and commercial users have also been little changed as the effect of a small growth in overall energy consumption has been offset by falls in coal consumption.

3.84 The UK will supply appropriate data to the European Commission under the monitoring decision of July 1993, continue to provide the IPCC/OECD with emissions data, and review longer term projections, taking account of developments since 1990.

CONCLUSION

3.85 The UK CO_2 Programme comprises a baseline inventory of emissions in 1990 compatible with IPCC methodology, a projection of emissions for the year 2000 and a range of measures aimed at returning emissions in 2000 to 1990 levels. Monitoring procedures are in place to check that the UK fulfils its commitments under the Convention.

NOTES

[1] *UK Emissions of Air Pollutants 1970-1990*. Gillham, C.A., Leech, P. K. and Eggleston, H. S. Warren Spring Laboratory, June 1992. ISBN 0-85624-747-2.

[2] *Digest of Environmental Protection and Water Statistics*. Warren Spring Laboratory. HMSO. 14th edition, 1991. ISBN 0-11-752634-7. 15th edition, 1992. ISBN 0-11-752815-3.

[3] *Digest of United Kingdom Energy Statistics 1993*. The Department of Trade and Industry. HMSO, 1993. ISBN 0-11-515328-4

[4] *This Common Inheritance: Britain's Environment Strategy*. HMSO Cm 1200, September 1990. *First Year Report*. HMSO, 1991. Cm 1655. *Second Year Report*. HMSO, 1992. Cm 2068.

CHAPTER 4

CARBON RESERVOIRS AND SINKS

CONVENTION COMMITMENTS

4.1 Fixing of CO_2 by trees followed by the storage of carbon in wood constitutes an effective reservoir for carbon. Soils are also large carbon reservoirs. The quantity of carbon contained in the four billion hectares of global forests is comparable with that currently in the atmosphere. The Convention recognises the importance of preserving and enhancing sinks and reservoirs of carbon, principally forests and soils, as part of a comprehensive approach to tackling climate change.

4.2 Article 4.1(b) requires national programmes to include measures to tackle climate change by addressing the removal by sinks of greenhouse gases, and Article 4.1(d) commits countries to promote sustainable management of biomass and forests. However, the specific commitment in the Convention to aim to return anthropogenic emissions of greenhouse gases to their 1990 levels by 2000 is not adjusted for sinks, as removals by sinks are referred to separately.

4.3 The UK intends to fulfil these commitments through a set of policies which will secure an annual increase in the net amount of carbon collected by sinks within the UK.

UK INVENTORY OF SINKS

4.4 The Convention commits all parties to prepare inventories of sinks. Scientific knowledge of sink processes is at present limited. The IPCC, with the support of the OECD, is working to prepare methodologies for inventories of sinks. There is a need to increase scientific understanding of the fixing of carbon by trees and soils, and the potential for increasing its impact on the limitation of CO_2 concentrations. In view of this uncertainty, the Government is sponsoring research at the UK Institute of Terrestrial Ecology (ITE) to improve our capacity to assess the role of carbon sinks (see Chapter Nine).

4.5 The UK has a detailed land classification system, and a soils database. These contain information which has been

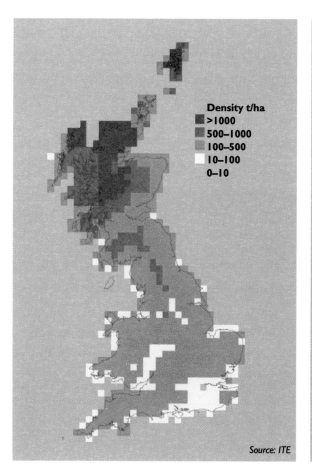

Figure 4a: Trend in carbon stored in soils across Great Britain[1].

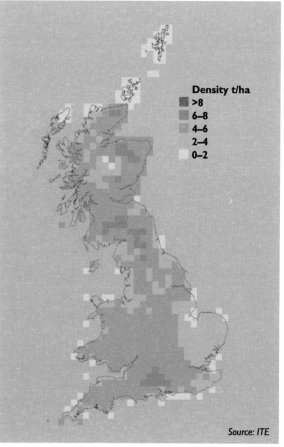

Figure 4b: Trend in carbon stored in vegetation across Great Britain[2].

used to quantify the present state of the vegetation and soil sinks, and to map their carbon density across Great Britain in 20 km by 20 km squares (see Figures 4a and 4b). The map in Figure 4b relies on the statistical relationship between land class (dependent on geology, climate, drainage, etc) and the vegetation likely to be encountered. Therefore, although the map shows the trend, it does not reflect particular features such as forests on the ground.

4.6 Typical amounts of carbon stored in UK soils range from 35 tonnes of carbon per hectare (tC/ha) for light arable cropland to more than 1000 tC/ha for deep peat. Carbon in vegetation generally falls between about 1 tC/ha for arable cropland and 70 tC/ha for woodland, though values up to about 200 tC/ha occur in some areas of mature woodland. However the statistical averaging process means that the trend in vegetation carbon density across Great Britain shows a much narrower range than this (see Figure 4b). The soil sinks covering the UK contain between 70 and 100 times as much carbon as the living vegetation. Peat dominates the soil sinks, and the vegetation sinks are predominantly forests. UK forests currently occupy about 10% of the land area, and in recent years the area covered by forests has been increasing by about 1% per year (see Figure 4c).

Figure 4c	UK carbon sinks in soil and vegetation[3]	
Reservoir	**Total carbon (MtC)**	**Main components**
Soil (including plant litter)	9,500 (range 7,500 - 11,500)	4,500 MtC in peat covering about 10% of land area; 5,000 MtC in other soils.
Vegetation	110 (range 90 - 130)	Mostly in woodlands covering 10% of land area.

Source: ITE

MONITORING OF CARBON SINKS

4.7 Soil and vegetation take time to adjust to changes in land use, so the current carbon densities shown in Figures 4a and 4b are not all equilibrium values. Notably, about 1 million of the 2.3 million hectares of forest is less than 40 years old and therefore still accumulating carbon. The present rate of carbon accumulation by new forest growth (including related litter and timber products) is estimated to be about 2.5 million tonnes a year, equivalent to about 1.5% of the UK's CO_2 emissions. The UK is assessing equilibrium carbon densities for soils and vegetation, so that long term trends in sink capacity can be estimated.

4.8 The UK will monitor how its policies are enhancing sinks by quantifying the uptake and loss of carbon. To do this the Government will use information on current and equilibrium carbon densities, and information about the dynamics of uptake and loss of carbon from soil and vegetation reservoirs in temperate countries. The impact of climate change on sink dynamics will be taken into account as far as it is practicable to do so.

4.9 Prior to the development of this monitoring system the UK has identified the main quantifiable human actions thought to have affected the UK's carbon uptake and loss in 1990 (see Figure 4d). Though the numbers are subject to considerable uncertainty, the calculations are consistent with the approach to sinks underlying the IPCC methodology (see Annex B).

4.10 The UK is undertaking research to improve the accuracy of the estimates in Figure 4d. Of the natural processes (which will nevertheless have been influenced by human activity), wetlands were accumulating carbon at an estimated 1.5 MtC/yr, but an estimated 1.1 MtC/yr was exported to the sea via river systems. More rapid photosynthesis induced by atmospheric nitrogen deposition and increased CO_2 concentrations in the atmosphere may have increased the soils and litter reservoir by 2 MtC, though this figure is very uncertain. In the long term an upward temperature trend would offset CO_2 fertilization by increasing the decay rates of soil organic matter. The UK is also conducting research in these areas.

MEASURES ON SINKS TO MEET CONVENTION COMMITMENTS

4.11 Despite the uncertainties, Figure 4d indicates that net sequestration of carbon was taking place in 1990 and the UK was therefore meeting its commitment to enhance sinks. In future years the UK will monitor how it is meeting its Convention commitment to protect and enhance sinks, taking into account the rate of recovery of sinks from past disturbances, and assessing the future impact of policies on forestry and land use.

Figure 4d Influence of land use changes on carbon uptake and loss by UK sinks[4]

Process	Carbon uptake (+) / loss (-) [a]
Agricultural land use change	0±0.5 MtC/yr
Afforestation [b]	+2.5 MtC/yr
Drainage of deep peats for afforestation	-0.2 MtC/yr
Drainage of lowland wetlands for agriculture	-0.2 MtC/yr
Peat extraction	-0.1 MtC/yr
Total	**+1.5 to +2.5 MtC/yr**

Note:
[a] Includes effects in 1990 of land use changes in previous years.
[b] Includes storage in trees, products and litter.

Source: ITE

Forestry

4.12 The UK is implementing forestry measures aimed at protecting its existing forests and steadily expanding tree cover. Tree cover in the UK has doubled this century. The UK is encouraging afforestation through incentive schemes such as the Woodland Grant Scheme and Farm Woodland Premium Scheme. These schemes are designed to encourage the creation of woodlands that provide a wide range of benefits – the production of timber, the enhancement of landscape and biodiversity, and additional opportunities for recreation in the countryside and close to towns and cities.

4.13 In the last decade afforestation in the UK has been at a rate of 20,000 to 30,000 hectares a year. The Government is making provision to spend £90 million supporting the expansion of tree cover over the three years 1994/95 to 1996/97, sufficient to assist the creation of some 60,000 hectares at present rates of grant of new woodlands and forests. The rate of carbon fixed by UK forests in trees, litter and products is likely to stay at about the present level of 2.5 MtC per year until 2000 (see Chapter Ten).

4.14 UK forestry policy is also aimed at the sustainable management of existing woodlands and forests. There is a general presumption against the conversion of forest land to other uses – the conversion of forest land to agriculture (the main cause of deforestation in the past) has virtually ceased. Measures are in place to encourage the positive management of forests and woodlands to enhance the many benefits that they can provide.

4.15 The UK's forestry policies are set out in detail in the UK Sustainable Forestry Programme.

Other Land use changes

4.16 There are a number of competing demands for land use in the UK: for agriculture, timber and minerals production; for housing, retail and industrial development and for roads; and for natural and semi-natural areas which provide important habitats for wildlife, the national heritage of landscapes and open spaces for recreation. The Sustainable Development Strategy[5] sets out how the Government seeks to balance these demands.

4.17 The principal land use change affecting 1990 carbon storage levels is the setting aside of arable land as a result of the recently reformed Common Agricultural Policy. Under the new rules, most farmers must set aside 15% of their arable land on a rotational basis, or 18% for 5 years or more. It is very difficult to predict the amount of land farmers will put into rotational or long term set-aside and, therefore, the effect it will have on UK carbon storage levels. However, in 1993 around 500,000 hectares changed from cultivated to non-cultivated land.

4.18 The effect on carbon storage will depend on the amount of time land remains uncultivated. With a 5 year set-aside period, there will be a significant increase in the carbon stored both above and below ground for the full five year period. However, if land is rotated on an annual basis, the effect on carbon storage levels may be low. Although new land is set aside each year, gains in stored carbon will be offset by ploughing land set aside in the previous year. The effect on carbon storage levels will also be influenced by the amount of set-aside land which is used to grow non-food crops (a permissible option under the scheme). Between 1988 and 1993, 130,000 to 155,000 hectares of land were in the former five year set-aside scheme (now closed).

4.19 The burning of agricultural straw was banned in England and Wales after the 1992 harvest. 3.5 million tonnes were burnt in 1992. However the available evidence suggests that the straw is likely to decompose fairly rapidly, and the soil carbon reservoir seems unlikely to increase by more than about 1 MtC overall, achieved over a few years. In addition the cessation of marine disposal may increase the amount of sewage sludge disposed of to land from about 0.5 million tonnes per year in 1990. Data on the dynamics of soil carbon from this source are very sparse, but the overall effect might conceivably be an increase of two or three MtC in the soil carbon sink, achieved over twenty or thirty years.

NOTES

1. The map, drawn on a 20km x 20km grid, is derived from the carbon content of the dominant soil type in each square kilometre of Great Britain, down to maximum depths of 1 metre in England and Wales and 4 metres in Scotland.
2. The map, drawn on a 20km x 20km grid, is based on the statistical relationship between vegetation and local geology, climate and drainage. Therefore the map shows trends, not particular features such as forests on the ground.
3. The table is consistent with the vegetation and soils maps in Figures 4a and 4b and assumes Scottish peat to have a bulk density of 110kgm^{-3}.
4. Uncertainties in the table are judged to be about ±20% for the afforestation entry and ±30% for the other entries, except agricultural land use change. See Annex B for underlying assumptions.
5. *Sustainable Development: the UK Strategy.* HMSO, 1994. Cm 2426. ISBN 0-10-124262-X.

CHAPTER 5

THE GREENHOUSE GAS PROGRAMME – METHANE

INTRODUCTION

5.1 Methane is the second most important anthropogenic greenhouse gas. Articles 4.1 and 4.2 of the Convention commit the UK to develop and publish national inventories of greenhouse gases and to take measures aimed at returning emissions of each greenhouse gas to 1990 levels by 2000. This Chapter describes the programme to meet those commitments for methane emissions: the methane programme.

5.2 The methane programme requires:
- an inventory of emissions in 1990 (the base year);
- an assessment of possible emissions trends to 2000 (the target year);
- a set of abatement measures designed to return emissions to 1990 levels by 2000; and
- a monitoring procedure.

METHANE EMISSIONS INVENTORY

5.3 UK anthropogenic methane emissions in 1990 are estimated to have been around 5 million tonnes (Mt). Figure 5a shows that the main sources are landfill waste (39%), agriculture (32%), coal mining (16%) and gas distribution (8%). Other sources include offshore oil and gas production (2%), sewage treatment and disposal (1%) and combustion processes and the petrochemical industries (2%).

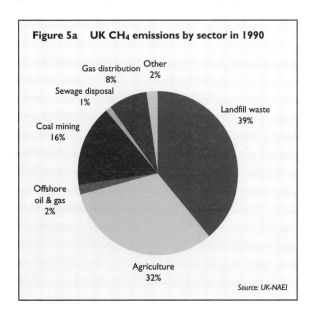

Figure 5a UK CH$_4$ emissions by sector in 1990
Source: UK-NAEI

5.4 There are considerable uncertainties in many of the factors used to calculate methane emissions for each sector. In 1991 the Government commissioned a review of the estimation methodologies for methane emissions. The review was carried out by a working group established by the Watt Committee on Energy, including experts from the relevant industries, from academic institutions and from Government. The report of this study, which includes an inventory of methane emissions, will be published shortly[1]. Error margins vary but remain wide in many cases, particularly for landfill gas.

5.5 New methologies were also put forward at a recent IPCC workshop and, for some sectors at least, appear to provide a means of estimating emissions more accurately. However, these have yet to be fully evaluated and incorporated into the national inventory. The emission estimates which form the basis of the programme have therefore been calculated on the basis of existing methodologies, though revised estimates using the new methodology are also referred to where appropriate in this Chapter. Annex B gives further details of the methods used to calculate emissions of methane from anthropogenic sources in the UK, and the uncertainties involved.

TRENDS IN EMISSIONS

5.6 As Chapter Two outlines, estimates of possible future methane emissions are subject to great uncertainty. However, for policy planning purposes, and based on the assumptions made for each of the individual sectors, the Government has adopted a working estimate of a 5% rise in overall methane emissions from 1990 to 2000 if no additional measures are taken to reduce emissions.

MEASURES TO LIMIT EMISSIONS

5.7 This is the first formulation of a comprehensive national strategy to reduce methane emissions in the UK. The

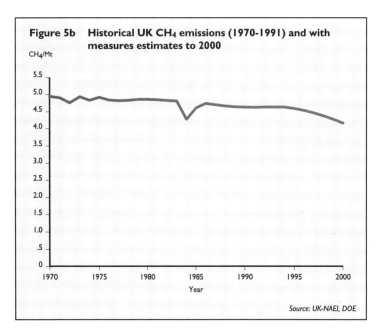

Figure 5b Historical UK CH$_4$ emissions (1970-1991) and with measures estimates to 2000
Source: UK-NAEI, DOE

measures in the programme should reduce the level of methane emissions by an estimated 0.6 Mt from 1990 to 2000, bringing emissions down to some 4.4 Mt, a reduction of just over 10% from the 1990 estimate of around 5 Mt (see Figure 5b). The UK would thus achieve more than the objective in Article 4.2 of the Convention.

Figure 5c   Anticipated changes in methane emissions from 1990 to 2000 by source

Source	Change in emissions 1990 to 2000 (Mt)
Landfill	-0.16
Agriculture	-0.1
Coal mining	-0.4
Gas distribution	-0.05
Oil and gas	+0.03
Sewage	+0.03
End use and petrochemicals	+0.01
TOTAL	-0.6

Note: Figures may not add up exactly due to rounding errors. There are considerable uncertainties in many of these estimates.

Source: DOE

5.8 Figure 5c shows that the largest reduction is anticipated in the coal mining sector. However, the most significant is for emissions from landfill which, without the measures in this programme, are assumed to rise by about 25% by 2000 (see paragraph 5.10). The net reduction in this sector is expected to be achieved in particular through the further encouragement of the utilisation of methane for energy generation, wherever practicable and cost effective. In other areas, the Government will encourage industry to draw up guidelines to limit methane emissions. The programme is comprehensive, covering all main sources and economic sectors, taking account of socio-economic factors. All measures in the programme are designed to be as cost effective as possible, utilising a mix of policy instruments and taking account of the circumstances of UK industry. The Government has closely involved industrial and other interested bodies in formulating the programme.

LANDFILL
Emission trends from landfill

5.9 Landfill sites are the largest single source of methane emissions in the UK, estimated in 1990 to be 1.9 Mt. Methane produced during the decomposition of waste may be conveniently divided into two categories: that arising from waste already in landfill sites, and that arising from the extra waste that is added to new and existing sites each year. Emissions were estimated by assessing the rate of decay for the different types of waste, and taking into account factors which either decrease the available carbon or remove methane prior to emission at the surface. These processes include aerobic decomposition, leachate carbon loss and methane oxidation.

5.10 Without measures, methane emissions are estimated to rise by about 25% from 1990 to some 2.5 Mt in 2000 for current planning policy purposes. The policies described in paragraphs 5.11 to 5.25 are expected to result in a decrease in emissions of about a quarter from the working estimate of 2.5 Mt to about 1.8 Mt in 2000.

5.11 The UK Government's policies are aimed at reducing the amount of methane from landfills by:
- adopting policies which promote waste minimisation and recycling, including energy recovery, and which reduce the amount of waste destined for final disposal; and
- introducing further measures to promote the use of methane from landfills as an energy source and the flaring of methane (which converts the carbon in methane to CO_2, a less powerful greenhouse gas).

5.12 The Government is also considering the use of economic instruments to correct any distortions in the waste disposal market, and so encourage waste destined for final disposal to be directed to the least environmentally sensitive disposal option (see paragraph 5.18).

Waste minimisation and recovery

5.13 The Government's policy is to minimise waste wherever practicable, but, where it is produced, to make productive use of it (either by recycling materials or by recovering energy) where possible.

WASTE MINIMISATION

The Government also runs programmes for industry offering advice on how to minimise waste, and supporting projects which demonstrate best practice. For instance the Government has supported a number of club projects in which companies take a collaborative and systematic approach to waste minimisation and publicise results at the end of the project. The DTI has published case studies on firms that undertook programmes of waste minimisation and found that their initial investment was very quickly repaid. The studies have shown that industry could reduce the amount of solid waste that it produces by up to 30% by adopting waste minimisation programmes.

5.14 The UK has taken a number of measures to minimise waste at source. The introduction of integrated pollution control under the Environmental Protection Act 1990[2] is designed to prevent or minimise all harmful releases to air, water, or land from the potentially most polluting processes.

5.15 The UK has a target of recycling half of all recyclable household waste by 2000. Policies being adopted to meet this target include:
- a scheme under which local authorities make payments to others in recognition of the savings made to waste disposal costs (recycling credits);
- a programme of capital investment for recycling by local authorities;
- grants to industry to develop new recycling technologies;
- a favourable legislative framework in the Environmental Protection Act 1990;
- a requirement on waste collection authorities to produce recycling plans; and
- the introduction of producer responsibility for waste, under which those whose products ultimately become waste are given responsibility for ensuring that the waste is used productively wherever practicable.

5.16 Government measures to raise landfill standards are expected to result in a rise in the cost of landfill over the rest of the decade (recent research for the Government suggests that the increase in costs may be between 35% and 135%). This is expected to increase the amount of waste which is recycled and incinerated.

WASTE TO ENERGY SCHEMES

Fifteen waste to energy schemes have been contracted to generate electricity under the first two NFFO orders on public electricity suppliers, although, due to planning and contracting problems, only seven are likely to be operating by 1995. Nevertheless, waste to energy is sufficiently economically attractive for it to have the potential to make a substantial contribution to the Government's aim of working towards 1500MW of new renewable electricity generating capacity by 2000. The Secretary of State for Scotland recently announced that he will introduce a Scottish Renewables Obligation (SRO), in parallel with the third round of the NFFO. This may lead to the development of waste to energy schemes in Scotland.

5.17 The Government is also promoting energy recovery from waste through orders requiring public electricity supply companies in England and Wales to obtain more electricity from renewable sources (see paragraphs 3.21 and 3.22). This initiative, known as the NFFO, has been operational since 1990. The NFFO has provided a guaranteed market and premium price for electricity generation from renewable sources sold to the public electricity supply companies.

5.18 One economic instrument being assessed (see paragraph 5.12) is the idea of introducing a levy on landfill. Depending on the size of the levy this could have a significant impact on the amount of waste going to landfill (see Figure 5d).

Figure 5d   Landfill levy - Estimated impact of levy set at different levels

Levy (£ per tonne)	Landfill	Incineration	Recycling
0	83	15	2
5	78	20	2
10	69	28	3
15	63	32	5
20	50	38	12

(% of household waste)

Source: Coopers & Lybrand

Measures to minimise methane emissions from individual landfill sites

5.19 Historically, legislation has dealt only with the public health dangers related principally to sub-surface migration of landfill gas. However, under the Environmental Protection Act 1990[2], the Government is implementing legislation which looks at harm to the whole environment. This should reduce methane emissions from landfill sites by minimising direct venting of gas to the atmosphere and increasing utilisation and flaring.

5.20 The Government will shortly introduce a new waste management licensing system to build on these measures. This is expected to impose still higher standards on landfill sites, which should increase the trend towards larger sites where the utilisation of landfill gas is economically viable as well as environmentally desirable.

> **NEW WASTE MANAGEMENT SYSTEM**
>
> Under the new waste management system, licences issued to landfill operators by waste regulation authorities will need to follow statutory guidance from the Secretary of State for the Environment. This guidance will be set out in Waste Management Paper 4 ("Licensing of Waste Management Facilities") and is expected to state that at present whole site landfill gas collection, followed by flaring or utilisation, should normally be installed in new sites, or existing sites with significant remaining capacity where active gas production is likely. The Government will fund further research relating to the control of methane emissions from landfills. It will also review the need for more guidance on the utilisation or flaring of landfill gas and on the design and operation of landfill sites to decrease net methane emissions, in the light of this research.

5.21 The exploitation of landfill gas as a renewable energy source is encouraged through the NFFO (see paragraph 5.17). Under the 1991 NFFO, the regional electricity companies have contracted with 28 projects for landfill gas generating stations to meet an obligation set at 48 MW. The third round of the NFFO includes a band for landfill gas from existing sites. The SRO will also include a band for landfill gas from existing sites, although the size of this band is still to be decided.

5.22 The Government runs two schemes whereby Supplementary Credit Approvals (permissions for a local authority to borrow funds for specific capital projects) are issued to local authorities in respect of landfill sites. The first, for action on contaminated land, includes work on closed landfill sites and is worth £12 million in 1993/94. The second covers both operational and closed sites, as well as by incinerators and other waste facilities, and is worth £13 million in 1993/94. Although the work is aimed principally at preventing the build up of methane for safety purposes, emissions are often reduced through improvements to, or the introduction of, pipes, ducts and flaring facilities. In Scotland, local authorities include proposals for capital expenditure on landfill gas in bids which the Government takes into account in the overall capital allocations to local authorities.

5.23 The Government is taking action to prevent the imposition of planning controls by local authorities that could otherwise limit the use of methane control measures on landfill sites. The Government has issued a new Planning Policy Guidance (PPG) Note on Planning and Pollution Control to local planning authorities, giving advice on the relationship between planning and the separate pollution control systems.

5.24 The Scottish Office Environment Department is preparing a National Planning Policy Guideline on land for waste disposal, which will also seek to avoid undue restrictions on landfill sites by ensuring that the planning system does not overlap with other relevant regulatory controls. A consultation draft National Planning Policy Guideline on Renewable Energy for Scotland has been issued.

> **PLANNING ADVICE AND LANDFILL SITES**
>
> The PPG on Planning and Pollution Control will advise that restoration conditions should not, as far as possible, restrict the use of pollution control measures that may be required on landfill sites. These may include measures to limit methane emissions either by flaring or by using it as a source of heat or electricity. The PPG also advises that consultation between the planning and pollution control authorities may be necessary to ensure that the pollution control measures do not unnecessarily prevent effective restoration of the site.

5.25 The Government has also issued a PPG Note on Renewable Energy which emphasises the benefits of renewables in terms of greenhouse gas emissions and draws attention to the significance of methane emissions.

AGRICULTURE
Emissions trends from agriculture

5.26 With emissions estimated at around 1.5 Mt in 1990, the agricultural sector is the second most important source of methane[3]. The majority of these emissions (68%) arise from enteric fermentation in ruminant livestock (cattle, sheep, goats, pigs, horses and poultry). Anaerobic digestion of livestock wastes is responsible for almost all of the remainder (31%). Crop residue burning is a very minor source (less than 1%) and, as the practice has been made illegal, with limited exceptions, from 1993, it is not considered further here.

5.27 Emissions from livestock and their wastes were calculated by applying Stage One IPCC/OECD methodology to livestock census data collected by the Ministry of Agriculture, Fisheries and Food (MAFF). If MAFF estimates for future livestock numbers are projected forward to 2000, then a 6% decrease in agricultural emissions is possible by that date.

Emissions from livestock and their wastes

5.28 The main reason for this projected decrease is the expected continued reduction in dairy cattle numbers (about 15% by 2000), a trend reinforced by recent reform of the CAP. Beyond this reduction in numbers, improving the productivity of individual animals offers the best possibility for reducing emissions direct from livestock themselves. In the short term at least, there is little room for improvement in this area as UK livestock are already some of the most efficient in the world. Research is underway to assess whether modified diets might lead to reduced emissions, but practical application of the results is unlikely to occur much before 2000.

5.29 The Government believes that, as in other sectors, farmers should minimise emissions from livestock wastes wherever this is practicable and cost effective. About half of all animal wastes are deposited in the field and little can be done about emissions from this source. For those wastes deposited in livestock buildings and subsequently stored prior to spreading on land, a number of possible means of reducing emissions have been suggested. However, considerable uncertainty remains about the effectiveness of these techniques and the costs are significant. MAFF has commissioned a £650,000 per year research programme to assess the various options. The results will be incorporated in advice to farmers and will inform future Government policy in this area.

COAL MINING
Emissions from coal mining

5.30 Coal mining was responsible for about 16% or about 0.8 Mt of UK methane emissions in 1990. The bulk of emissions from the coal industry are from deep mines and British Coal has carried out an extensive monitoring and analysis programme to quantify the methane emissions from coal extracted from deep mines in the UK. The final emission values take into account losses of methane during mining, transportation and storage of coal.

5.31 Coal production in the UK has been declining for many years and this process is expected to continue to some extent. The Government has adopted a working assumption that methane emissions from this source could fall by about a half to around 0.4 Mt by 2000.

Measures to Limit Emissions

5.32 The Government is encouraging operators to increase the proportion of methane from mines which is utilised, and will take steps in the forthcoming privatisation of the coal industry to require operators to publish periodic statements of their approach to limiting methane emissions. It is envisaged that the statement would describe estimated emissions, measures being taken or proposed to limit emissions and their effect, monitoring procedures, and research, where appropriate, into abatement methods.

5.33 The Government recently commissioned a study to evaluate the technical potential, cost and effectiveness of possible measures to limit methane emissions from UK deep-mined coal production. In particular, the study is considering means to limit emissions which could be taken by the industry before 2000. It is also considering longer-term measures which might limit emissions beyond 2000. Industry is expected to use the report as a basis for developing practical guidelines on the control of methane emissions wherever this is practicable and cost-effective.

GAS DISTRIBUTION SYSTEM
Emission trends from transmission and distribution of natural gas

5.34 UK methane emissions from natural gas transmission and distribution are estimated to have been some 0.4 Mt in 1990. This was derived from a recent major programme of measurement by British Gas of leakage rates, in which they considered mains and services combined and separately. If no action were taken to prevent leakage, emissions would rise slowly in future. However, the leakage control strategy of British Gas should lead to a decrease in emissions of around 15%, moving closer to 0.3Mt by 2000.

Leakage control strategy

5.35 In 1992, British Gas established, following their comprehensive leakage test programme, that leakage amounted to about 1% of gas throughput. The majority of the leakage comes from the low pressure distribution system. The high pressure gas system (including supplies to power stations) is routinely almost leak free, with some venting at pressure release valves and routine venting at pilot operated reduction and constant pressure valves.

5.36 British Gas has reviewed its leakage control policy and will be taking action to control gas leakage further through the extended use and enhancement of gas conditioning and pressure management.

5.37 Gas leakage in the low pressure system is also being reduced as the existing system is being replaced or refurbished. Mains replacement is not a cost effective leakage reduction technique and is primarily determined by safety considerations. However, the need to reduce overall leakage rates will continue to be a criterion for deciding replacement priorities. A total of 2,400 km per year is provisionally planned up to 2000. In addition, new mains are currently being laid at a rate of 4,200 km per year with 315,000 new services added and the same number replaced each year. All newly laid and replacement mains will be of new polyethylene which has a leakage rate significantly lower than that of traditional steel and iron mains.

5.38 Overall, British Gas is provisionally targeting to reduce leakage rates by approximately 2% a year up to 2000 by the complementary use of gas conditioning, pressure management and replacement. The finalised leakage reduction strategy will include target emissions reductions and implementation and investment plans for the various leakage reduction techniques. British Gas is developing details of this strategy, which will not be finalised until the outcome of an inquiry by the Monopolies and Mergers Commission is known.

OIL AND GAS PRODUCTION
Emission trends from oil and gas production

5.39 Methane emissions from oil and gas production were estimated to be around 0.1 Mt in 1990. These estimates were derived by the United Kingdom Offshore Operators Association (UKOOA) from data supplied by operators on emissions in 1991. These showed that the main sources of emissions were venting of gas to the atmosphere for emergencies or planned maintenance, unburnt gas from flares, and fugitive emissions from valves and other components.

5.40 Oil and gas production is expected to increase over the next few years, but current expectations are that it will peak before the end of the century. Estimated output is not necessarily a good indicator of likely methane emissions because emissions are more closely related to field specific operational requirements. The UK Government has adopted a working assumption of a 30% increase in emissions to 0.13 Mt by 2000.

5.41 There is increasing interest in the UK in the possibility of commercial exploitation of coal deposits as sources of methane. A number of companies showed interest in coal bed methane in the last round of landward petroleum licensing. The technology envisaged for such developments is broadly similar to that used for other onshore oil and gas developments.

Measures to reduce emissions

5.42 Following consultations with the industry, the Government commissioned an independent study in March 1993 to assess the options available for limiting methane emissions from the oil and gas production industry.

STUDY INTO THE METHANE EMISSIONS FROM OIL AND GAS PRODUCTION

The aim of the oil and gas production industry study was to evaluate the technical potential, cost and effectiveness of possible measures to limit emissions. The study looked particularly at the means which could be taken by the industry to limit emissions by the year 2000, but also the longer term beyond 2000. The study concluded that there was technical scope for reducing emissions, although the cost effectiveness of measures needed to be considered in more detail. A number of options for reducing emissions were suggested for further consideration. The industry is committed to good environmental practice. It wil use the report as a basis for developing practical guidelines on the control of methane emissions, wherever this is practicable and cost effective.

SEWAGE TREATMENT AND DISPOSAL
Emissions from sewage treatment and disposal

5.43 Most sewage produced in the UK is treated before disposal. Treatment is by anaerobic digestion, following which the sludge is usually left to settle in open tanks. The sewage sludge is then disposed of by one of four methods. Currently some 1 Mt of dry solid is disposed of annually, with 47% going to the land, 30% dumped at sea, 17% used as landfill and 6% incinerated. Methane is emitted from digesters as a result of emergency venting, leaks and unburnt gas in flares, from settlement tanks and from the disposal of sludge. Overall, methane emissions from this source (excluding emissions from sewage used as landfill) have been estimated at around 0.07 Mt in 1990.

5.44 Emissions are likely to rise in the future as a result of the implementation of the EC Urban Waste Water Treatment Directive[4], but the rate of increase will depend upon any change in the relative importance of the different sludge disposal routes. The UK will stop dumping sewage at sea by the end of 1998. Using the emission factors reported by the Watt Committee and assuming no reduction measures, future emissions of methane are predicted to increase by around 45% to some 0.1 Mt by 2000.

Measures to reduce emissions

5.45 It is difficult to assess current best practice for limiting methane emissions from the sewage industry, as processes and procedures are largely driven and constrained by local circumstances. Nevertheless, the Water Services Association believes that the changes in sludge utilisation practise brought about by changes in legislation and a general updating of sludge treatment practices will result in the vast majority of sludge being treated and utilised in ways which limit methane production without entailing excessive cost. The Government will encourage further studies in this area, with a view to developing guidelines on best practice to limit methane emissions wherever this is practicable and cost-effective.

COMBUSTION PROCESSES AND PETROCHEMICAL INDUSTRIES

Emissions trends from combustion processes and petrochemical industries

5.46 The Watt Committee included two other minor categories in its report: end use and combustion processes and petrochemical and process industries. End use and combustion processes include stationary and mobile (road transport) combustion sources and leakage of natural gas from customer appliances. Emissions in 1990 have been estimated to be under 0.1 Mt. The working assumption is that emissions could rise by around 15% by 2000.

5.47 Emissions from petrochemical and process industries are made up largely of losses from refineries and industrial processes. Emissions in 1990 are estimated to be about 0.01 Mt. The working assumption is that emissions from this source may rise by up to 10% by 2000.

Measures to reduce emissions

5.48 Emissions from road transport will be reduced by the introduction of catalytic convertors. Under the EC Directive 91/441/EEC[5] all new petrol engined cars sold in the UK must now be fitted with three way catalysts.

5.49 Oil refining is a controlled process under the Environmental Protection Act. Her Majesty's Inspectorate of Pollution (HMIP) regulate emissions from controlled processes and ensure that emissions are as low as reasonably achievable using BATNEEC.

RESEARCH

5.50 Over the last 5 years, the Government's Department of the Environment has funded 28 research projects related to landfill gas worth almost £5 million. Although several are aimed principally at controlled waste management, they will increase our understanding of controlling methane emissions. Projects include an assessment of methane oxidation in soils to try to inhibit methane production in landfills, a study of the effectiveness of remedial control measures for landfill sites, measurement of gas flux around landfill sites and a review of the contribution of landfilling to greenhouse gas emissions. A £650,000 a year research programme assessing the possibilities for reducing emissions from farmed livestock and their wastes is currently being undertaken by MAFF.

> **BIOFUELS RESEARCH**
>
> The DTI's Biofuels Programme aims to encourage the widespread commercial exploitation of the resource by identifying and demonstrating those areas in which it could be competitive and environmentally acceptable. Developments in Biological Processing, which is one important element of the Biofuels Programme, include good progress in fundamental landfill macrobiology, landfill gas field studies and the anaerobic digestion programme. On the international front, the Biofuels Programme has been strengthened by collaborative activities, particularly on landfill gas and biological processing of waste. On the European front, the Government will seek to establish closer ties with the Community programmes to derive maximum benefit from other related activities.

MONITORING

5.51 There are significant uncertainties in many of the factors used to calculate emissions for each sector and the final emission estimates are subject to wide error margins. These emission estimates will be revised as further advice is obtained from government and industrial organisations, and results of research programmes become available. In particular, the level of emissions from landfill is still subject to large uncertainties.

5.52 The Government will ensure that emitters will take steps to monitor emissions, particularly the coal, oil, gas and landfill industries. In agriculture, the diffuse sources of methane, such as ruminant animals and animal wastes, makes recording of emissions much more problematic, but the Government has put in place a research programme which should result in these emissions being quantified more accurately in future. Estimates of methane emissions, using the best available data will be published annually.

5.53 The programme will need to be kept under review in the light of updated information, as part of the monitoring process, so that the UK meets its Convention commitments.

Like CO_2, short term deviations from the trends implied by the working assumptions are inevitable, and would not necessarily imply that the programme's measures needed to be changed unless they indicated a change in the longer term.

CONCLUSION

5.54 The UK's methane programme is based on an inventory consistent with current IPCC methodologies, and includes an assessment of possible trends in emissions to 2000 and a set of measures aimed at reducing emissions. It is expected that this programme should lead to a 0.6 Mt (or just over 10%) reduction below the 1990 levels of around 5 Mt, which would mean emissions in 2000 would be around 4.4 Mt. Procedures are in place to improve emissions estimates and assessment of possible future emissions trends, and to monitor progress over the period to 2000.

NOTES

1. *UK Methane Emissions.* Watt Committee on Energy, London, in press.
2. *The Environmental Protection Act,* 1990. HMSO, 1990. ISBN 0-10-544390-5.
3. A new methodology for calculating agricultural emissions has been put forward very recently (see paragraph 5.5), which the UK believes produces more accurate estimates, in particular for emissions from livestock wastes. Using this methodology, agricultural emissions are estimated at around 1.2 Mt in 1990, with by far the most significant source being enteric fermentation (89%), followed by livestock wastes (9%) and crop residue burning (2%). On the same basis, agricultural emissions are expected to fall by 7% by 2000.
4. *EC Urban Waste Water Treatment Directive.* Directive no. 91/271/EEC.
5. *EC Vehicle Emissions (Amendment) Directive.* Directive no 91/441/EEC.

CHAPTER 6

THE GREENHOUSE GAS PROGRAMME – OTHER GREENHOUSE GASES

NITROUS OXIDE

6.1 Nitrous oxide is about 270 times more powerful as a greenhouse gas than CO_2 over a one hundred year time horizon. It is released in much smaller quantities than CO_2 and contributes, at current estimates, around 6% of the direct global warming effect. There is, however, some uncertainty as to whether the inclusion of indirect effects would increase or decrease this figure. As for other greenhouse gases, the Government is committed to action aimed at returning nitrous oxide emissions to 1990 levels by 2000.

Emissions inventory and trends

6.2 Total anthropogenic emissions of nitrous oxide in the UK are estimated to have been around 0.11 Mt in 1990, with the major sources being the manufacture of nylon and agriculture (see Figure 6a).

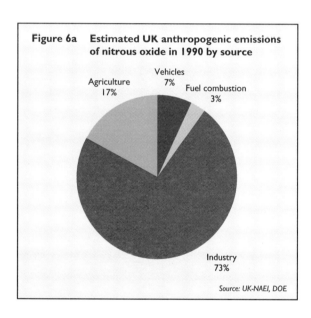

6.3 There are considerable uncertainties in many of the factors used to calculate emissions of nitrous oxide. Current emission estimates and projections to 2000 are based on the latest information available on agricultural practices, industrial processes and road transport policies in the UK. As for methane (see paragraph 5.5), existing methodologies have been used for this programme, though revised estimates using recently proposed methodologies are also referred to in this Chapter where appropriate. Details are given in Annex B.

6.4 The projections include the impact of the control measures that the Government and industry will be adopting in the near future. If no new measures were taken by Government or industry, nitrous oxide emissions would be expected to fall slightly. However, with these measures (see paragraphs 6.5 to 6.15), the UK expects that annual emissions of nitrous oxide should fall substantially from around 0.11 Mt 1990 to around 0.03 Mt in 2000 (see Figure 6b).

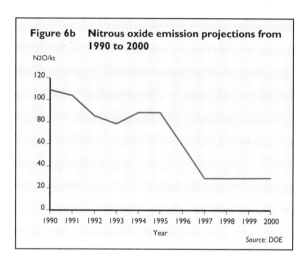

Measures to limit emissions

Nylon manufacture

6.5 Nitrous oxide is emitted during the manufacture of adipic acid, which is an intermediate in the production of nylon. Emission levels depend on the level of production of adipic acid, which in turn fluctuates with market demand for nylon. Emissions in 1990, when there were two adipic acid plants in the UK, were an estimated 80 kt. Emissions dropped to around 57 kt in 1992 when market demand fell and one plant was partially shut down, leaving about two thirds of the original capacity still operating. Maximum annual production at present would result in nitrous oxide emissions of about 60 kt.

6.6 However, since nitrous oxide was recognised as a greenhouse gas, a number of major adipic acid producers have been investigating different abatement strategies. Producers formed an inter-industry group to share information on technologies being developed for nitrous oxide control. These include incineration, conversion to recoverable nitrogen monoxide and catalytic decomposition to nitrogen and oxygen.

6.7 DuPont (U.K.) Ltd, the owners of the plants in the UK, are committed to minimising nitrous oxide emissions subject to BATNEEC, using the catalytic destruction option by mid-1996. Trials to determine the best catalyst and operating conditions have shown that emissions of nitrous oxide will fall to less than 0.02 kt each year, once an abatement system has been fitted. There may be slight fluctuations (ranging from

0.01 to 0.04 kt) depending on the operating conditions, the extent of recycling of gas, and the type and condition of the catalyst. This extremely low emission rate of nitrous oxide will remain constant after 1996 assuming the demand for adipic acid remains at current levels.

6.8 Nitrous oxide is a prescribed substance under the Government's Integrated Pollution Control (IPC) system, which was introduced to regulate industrial pollution under the 1990 Environmental Protection Act (see paragraph 5.14), and its emission by nylon production plants will therefore be regulated by HMIP. When considering emission authorisations, HMIP will ensure that emissions from the process are as low as can be reasonably achieved through the use of BATNEEC. DuPont submitted an application for an authorisation under IPC in October 1993. This included the proposed abatement strategies outlined in paragraph 6.7. The application is currently under determination.

Agriculture

6.9 Soils give rise to emissions of nitrous oxide as part of the natural process of nitrogen circulating within the environment. Various agricultural practices can, however, provide additional sources of nitrogen to the soil and therefore increase nitrous oxide emissions.

6.10 Using an OECD-recommended methodology, the Government estimates that just over 4 kt of nitrous oxide emissions resulted from the application of chemical fertiliser to land in 1990. Estimated emissions in 1990 are around 14 kt from the spreading of livestock wastes and just 0.2 kt from crop residue burning. The UK therefore estimates total emissions from agriculture to be about 18 kt. However, there are considerable uncertainties in all of these calculations[1].

6.11 Fertiliser use is currently declining and the underlying pressures of the CAP reform should reinforce that trend. Significant factors influencing the trend include reduced cereal prices and a more than threefold expansion to 500,000 hectares of set-aside land, which will be largely unfertilised. Fertiliser use should decline by about 10% by 2000, leading to a similar decline in emissions of nitrous oxide from this source. Emissions from livestock wastes should also fall, based on predictions of livestock numbers up to 2000. Crop residue burning was, with limited exceptions, banned in 1993. It is therefore estimated that there will be an 8% decrease from 1990 levels of emissions from agriculture by 2000.

6.12 The Government believes that, as in other sectors, farmers should minimise emissions of nitrous oxide wherever this is practicable and cost-effective. However, on the basis of current knowledge, it appears that the scope for reductions from this sector beyond those resulting from lower livestock numbers or fertiliser use is very limited. To help expand knowledge in this area, the Government is funding a significant research programme, including:

- research on the collection of emissions data. The Government should then be able to identify best practice and accurately forecast the effects;
- work aimed at assessing the effects of changing fertiliser use;
- assessment of the use of substances to inhibit the amount of nitrous oxide released from fertilisers; and
- studies on the control of nitrous oxide from livestock wastes.

Future research will look at changes in fertiliser application methods to reduce emissions without at the same time increasing nitrate leaching.

Motor vehicles

6.13 Road transport accounted for a relatively small percentage (7%) of the national emissions of nitrous oxide in 1990. Emission estimates for this source are based on the methodology used in the UK-NAEI to assess all transport-related emissions, and the emission factors are in line with current IPCC/OECD recommendations. Projections suggest that emissions of nitrous oxide from vehicles in the UK will increase by around 17% from just under 8 kt a year in 1990 to around 9 kt a year in 2000.

6.14 These emissions are projected to rise by 2000, mainly because of the required fitting of 3-way catalysts on all new petrol-engined cars. This reduces the emissions of other oxides of nitrogen by 75% (see paragraph 6.24), which should bring a significant improvement in urban air quality, but a side effect is that nitrous oxide is produced in the catalytic process. Measurements made on the exhaust emissions of cars fitted with catalytic converters show that emissions of nitrous oxide are a factor of 3 to 5 times higher.

6.15 Although the reason for nitrous oxide formation is not fully clear, poor catalyst efficiency may contribute to excess emissions. The Government will ensure that inspection and maintenance programmes and catalyst durability are carefully controlled to ensure efficient catalytic performance over a vehicle's lifetime.

TROPOSPHERIC OZONE

6.16 Ozone is an effective greenhouse gas both at lower levels in the atmosphere (tropospheric ozone) and at higher altitudes (stratospheric ozone). Unlike other major greenhouse gases, direct emissions of ozone are negligible.

Low-level ozone is formed mainly through a series of complex reactions driven by sunlight and involving nitrogen oxides, volatile organic compounds (VOCs) and carbon monoxide. Current best estimates suggest that low-level ozone has caused up to 10% of the total global radiative forcing since pre-industrial times.

6.17 Nitrogen oxides, VOCs and carbon monoxide have negligible direct greenhouse effects. However, as these gases contribute to the photochemistry which influences tropospheric levels of more powerful greenhouse gases such as ozone, and to a lesser extent methane, hydrochlorofluorocarbons and hydrofluorocarbons, they contribute *indirectly* to the global radiative forcing. Current limitations in the understanding of the photochemical processes involved mean that these indirect effects cannot be accurately quantified at present, but they are likely to be significant.

6.18 Control strategies must be aimed at reducing the precursors to ozone formation. Nitrogen oxides, VOCs and carbon monoxide are all emitted directly into the atmosphere as a result of human activities, and the majority of policies to reduce low-level ozone levels are concerned with reducing emissions of these pollutants. These are considered in the following sections.

NITROGEN OXIDES

6.19 Nitrogen oxides are formed during the combustion of fossil fuels and contribute to the greenhouse effect as a precursor to low-level ozone. At the same time, however, they lead to an increase in the oxidising capacity of the atmosphere and so a decrease in the concentration of methane. Thus, enhanced levels of nitrogen oxides have opposing effects on the abundance of two greenhouse gases, ozone and methane, and it is possible that the overall effect may be beneficial. But the magnitude of these indirect greenhouse effects remains highly uncertain and a reliable estimate of the overall radiative forcing of nitrogen oxides cannot be made at present. It is likely that the relatively low emissions mean that nitrogen oxides have only a minor influence on the overall greenhouse effect.

Emissions inventory and trends

6.20 The main anthropogenic sources of nitrogen oxides are road transport and power stations (see Figure 6c) with total UK emissions of nitrogen oxides from anthropogenic sources in 1990 estimated to have been 2.8 Mt (expressed as NO_2 equivalents).

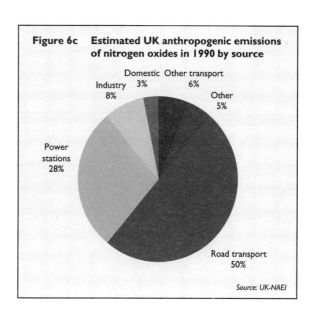

Figure 6c Estimated UK anthropogenic emissions of nitrogen oxides in 1990 by source

Source: UK-NAEI

6.21 Emissions projections for nitrogen oxides are based on the methodology used in the UK-NAEI for historical estimates, and extended to take account of possible future trends in energy use and assumptions about technology, as well as the effects of national policy requirements. As the main source of nitrogen oxides is fuel combustion, assumptions about national fuel use are crucial to any projection of future emissions.

6.22 Estimates show a steady decline in anthropogenic emissions of nitrogen oxides to between 1.8 Mt a year and 2.2 Mt a year by 2000 (see Figure 6d), largely as a result of existing policy measures (see paragraphs 6.23 to 6.25). The uncertainties in estimating total UK emissions of nitrogen

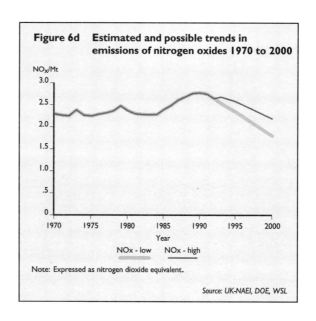

Figure 6d Estimated and possible trends in emissions of nitrogen oxides 1970 to 2000

Note: Expressed as nitrogen dioxide equivalent.

Source: UK-NAEI, DOE, WSL

oxides from anthropogenic sources may be less than for some greenhouse gases, although significant uncertainties still exist with respect to the effect of catalysts and driving conditions on emissions from vehicles, and to many of the assumptions related to emissions from power stations.

Measures to limit emissions

6.23 The UK is committed to reducing emissions of nitrogen oxides under the EC's Large Combustion Plants Directive (LCPD)[2]. Taking 1980 as the baseline, the UK must reduce nitrogen oxides from large plants by 15% by 1993 and 30% by 1998. A national programme and plan published in 1990 sets out the UK's annual quotas for power stations, refineries and other industries necessary to achieve these reductions, and these will be met chiefly by the fitting of low-nitrogen oxides burners to existing plants. The implementation of the Directive will contribute to the UK's commitment under the Convention, with emissions from the three sectors anticipated to fall to significantly less than 1990 levels by 2000.

6.24 In addition to the LCPD, there are EC Directives to control nitrogen oxides emissions from road vehicles. Since the end of 1992, new petrol driven cars must be fitted with catalytic converters to meet new EC emission standards. As a result, emissions of most pollutants, including nitrogen oxides, volatile organic compounds, and carbon monoxide, will be reduced by up to 80% for each new car sold. The EC has also agreed new, tighter diesel standards for buses and lorries, which will be implemented in two phases by 1996–97. As a result, emissions of nitrogen oxides will be reduced by typically 50% for each new bus or lorry.

6.25 The UK is also committed to a protocol signed in 1988 within the United Nations Economic Commission for Europe (UNECE) to return nitrogen oxides emissions from all sources to 1987 levels by 1994[3]. The UK is on course to comply with this commitment. In the longer term it is estimated that by 2005 nitrogen oxides emissions should be around 30% down on 1987 levels.

VOLATILE ORGANIC COMPOUNDS

6.26 "Volatile Organic Compounds" (VOCs) represent in this report all organic compounds, other than methane, that are capable of producing photochemical oxidants, such as ozone in the atmosphere, by reaction with nitrogen oxides in the presence of sunlight. Methane is considered separately in Chapter Five.

6.27 VOCs have negligible direct greenhouse effects, but have positive indirect effects through their role in the formation of tropospheric ozone. In addition, increased levels of VOCs reduce ambient concentrations of atmospheric oxidants thus leading ultimately to enhanced levels of the potent greenhouse gases methane, hydrochlorofluorocarbons and hydrofluorocarbons. Current uncertainties in the detailed understanding the role of VOCs in tropospheric chemistry prevent an estimation of the magnitude of this positive indirect greenhouse effect.

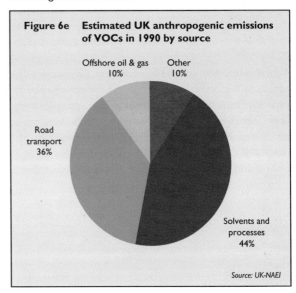

Figure 6e Estimated UK anthropogenic emissions of VOCs in 1990 by source

Offshore oil & gas 10%
Other 10%
Road transport 36%
Solvents and processes 44%

Source: UK-NAEI

Emissions inventory and trends

6.28 VOCs are emitted from a wide variety of sources including road transport, the use of solvents and industrial processes. Total VOC emissions in 1990 from anthropogenic sources in the UK were approximately 2.7 Mt (see Figure 6e).

6.29 Projected emissions for VOCs are given in a strategy document published recently by the Government, "Reducing Emissions of VOCs and Levels of Ground-level Ozone: a UK Strategy"[4]. This report provides details of anticipated future VOC emissions by sector, taking into account existing and anticipated abatement policies, and sets out the UK response to meeting its obligations under the 1991 UNECE protocol on control of VOC emissions[5]. Projections show a decrease in emissions to around 1.7 Mt by 2000.

Measures to limit emissions

6.30 The main measures to reduce emissions of VOCs are the EC Directive requiring catalytic convertors on new cars from the end of 1992 (see paragraph 6.24) and controls on industry under the Environmental Protection Act 1990. The EC Directive also requires cars to be fitted with a small carbon canister to control evaporative emissions. VOCs are prescribed substances under the 1990

Act, and industrial operators must apply for authorisation for any emissions. The pollution control authorities will lay down conditions of operation, including emission standards based on BATNEEC. In addition, forthcoming EC directives are expected to control VOC emissions from solvent-using industries, and from the storage and distribution of petrol. As a result of these measures it is estimated that emissions of VOCs will be reduced by over 35% from 1990 levels by 2000 (see Figure 6f).

CARBON MONOXIDE

6.31 Carbon monoxide affects the radiation budget of the atmosphere in a similar way to VOCs. The photochemical processes involving carbon monoxide and nitrogen oxides lead to changes in the concentrations of low-level ozone and atmospheric oxidants, so that high levels of carbon monoxide result in enhanced concentrations of both ozone and methane, and hence a positive indirect greenhouse effect.

Emissions inventory and trends

6.32 In the UK, emissions of carbon monoxide from road transport accounted for approximately 90% of the total anthropogenic emissions in 1990 of 6.7 Mt (see Figure 6g).

6.33 Future emissions are expected to fall to between 3.4 Mt a year and 3.7 Mt a year by 2000 (see Figure 6h). The methodology used to estimate vehicle emissions for the UKNAEI is well-established and continually updated as new information and improved emissions factors for vehicles become available.

Measures to limit emissions

6.34 As stated in paragraph 6.24, the installation of three-way catalysts since the end of 1992 will rapidly reduce emissions of carbon monoxide from vehicles by around 70% by 2000, and therefore total UK emissions are also projected to fall substantially. It is estimated that total UK carbon monoxide emissions will be reduced by up to 50% from 1990 levels by 2000.

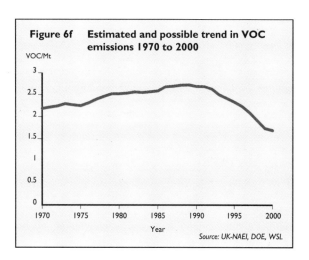

Figure 6f Estimated and possible trend in VOC emissions 1970 to 2000

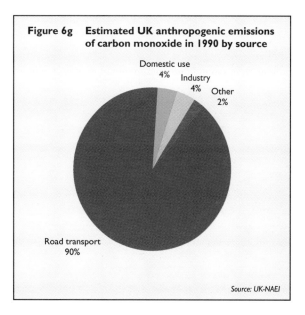

Figure 6g Estimated UK anthropogenic emissions of carbon monoxide in 1990 by source

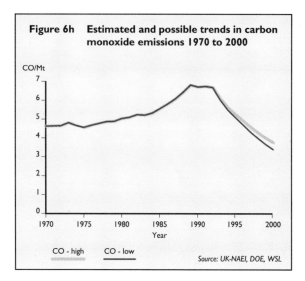

Figure 6h Estimated and possible trends in carbon monoxide emissions 1970 to 2000

HALOCARBONS

Chlorofluorocarbons, Halons, Carbon Tetrachloride, Methylchloroform, Hydrochlorofluorocarbons (HCFCs) and Hydrobromofluorocarbons

6.35 A number of greenhouse gases are also ozone depleting substances controlled under the Montreal Protocol. These substances are excluded from the requirements of the Convention's commitments, and this programme does not, therefore, provide inventories for them, nor details of the measures being taken to limit their emissions. It is worth noting, however, that the UK, along with its EC partners, will phase out the production and use of CFCs, halons and carbon tetrachloride more rapidly than the Montreal Protocol requires.

6.36 In addition the UK has legislated to control emissions of ozone depleting substances, by making it illegal to dispose of waste ozone depleting substances in a manner likely to cause pollution to the environment. A regulation to control consumption of HCFCs is currently under discussion in the EC. Again this will impose tighter controls than those set out in the Protocol, including use controls which will strictly limit emissive uses of HCFCs.

Hydrofluorocarbons (HFCs)

6.37 HFCs are important alternatives for some uses of CFCs and HCFCs. They are not ozone-depleting gases and are not therefore covered by the Montreal Protocol. But they are greenhouse gases, and the UK is committed under the Convention to take measures aimed at returning emissions of HFCs to 1990 levels by 2000. The Government has already taken steps in this direction in that the controls which have been introduced on the disposal of waste ozone-depleting substances (see paragraph 6.36) also apply to waste HFCs.

6.38 However, there are difficulties in achieving the UK's commitment in full. Since HFCs were not produced or used in significant quantities in the UK in 1990, production is bound to increase between then and 2000. The Government will therefore be exploring further with HFC producers and consumers how best to make progress in relation to the UK's commitment on HFCs without damaging current efforts to phase out the ozone-depleting substances which they replace. In particular, it will be exploring the scope for voluntary agreements to ensure that, where HFCs are used, emissions are minimised and that HFCs are not used where emissions are unavoidable if safe, practical and more environmentally acceptable alternatives are available.

6.39 One HFC (HFC-23) is produced as a by-product during the manufacture of HCFC-22, which is an HCFC that will be controlled under the Montreal Protocol. At present HFC-23 constitutes between 2 and 4% of HCFC-22 production, and is released to the atmosphere. The Government is exploring the scope for achieving significant reductions in HFC-23 emissions with UK producers of HCFC-22.

PERFLUOROCARBONS (PFCS)

6.40 PFCs are powerful greenhouse gases with very long atmospheric lifetimes. Some of the higher PFCs may be needed as replacements for a few fire fighting applications where other substitutes for halons are not available. They are also used in a few contained solvent systems for delicate high value instruments. These uses will not give rise to significant emissions, not least because of the high price of these PFCs.

Carbon tetrafluoride (CF_4) and hexafluoroethane (C_2F_6)

6.41 CF_4 and C_2F_6 are perfluorocarbons that are extremely strong greenhouse gases. It appears that the primary reduction of aluminium is the only source of UK emissions of these PFCs, and the aluminium industry has commissioned several reports into the fate and effect of PFCs in the upper atmosphere and the sources and rates of emissions.

6.42 The accurate measurement of emission rates of CF_4 and C_2F_6 has proved difficult, and, due to this uncertainty and the small size and diversity of the UK aluminium industry, no measurements of emission rates have been made in the UK. However, it appears that emissions of these PFCs are confined to the period that an aluminium reduction cell undergoes an "anode effect". Using emission factors calculated by the US Environmental Protection Agency and actual and projected number and duration of anode effects, the estimated UK emissions from 1990 to date, and also estimated trends to 2000 are shown in Figure 6i.

6.43 The reductions in estimated emissions since 1990 are due to action taken by the two companies who operate the four UK primary aluminium smelters. The management of each UK primary aluminium smelter has put in place plans to reduce the frequency of anode effects at their location. They have directed efforts towards improvements in operation and process controls, particularly the development of automatic process control systems installed in three of the four smelters.

6.44 As Figure 6i shows, results have so far been encouraging, with a 70% reduction in CF_4 and a 65% fall in C_2F_6 over the last three years. UK industry has identified development routes that should lead to further reductions, and is confident that the good progress to date can be continued to reduce emissions of both CF_4 and C_2F_6 to more than 90% below 1990 levels by 2000.

6.45 The International Primary Aluminium Industry are organising a conference in London on PFCs in March 1994 for industry representatives and researchers, which will aim to determine a world wide strategy of measurement and control of these compounds.

6.46 CF_4 and C_2F_6 are prescribed substances under the Environmental Protection Act 1990, and their release is regulated by HMIP. The aluminium industry and HMIP have regular discussions about the release of these PFCs, and the companies will be applying to HMIP for authorisation of the releases in 1995. When determining authorisations HMIP ensures that BATNEEC are used to prevent, minimise or render harmless the release of prescribed substances.

Figure 6i Estimated emissions of CF_4 and C_2F_6 from the UK primary aluminium smelting industry for 1990 to 1993 and projections for 2000

	1990	1991	1992	1993[1]	2000[2] (high)	2000[2] (low)
CF_4/t yr^{-1}	274	213	102	80	31	10
C_2F_6/t yr^{-1}	28	21	10	8	3	1

Note: [1] Figures for 1993 have been extrapolated to 12 months.
[2] The high and low forecasts for 2000 represent two possible scenarios for primary production capacity in the UK. The "low" figure is considered to be more likely.

Source: ALFED

NOTES

[1] A new methodology for calculating agricultural emissions has been put forward very recently (see paragraph 6.3) which the UK believes produces more accurate estimates, in particular for emissions from use of chemical fertilisers and from livestock wastes. Using this methodology, agricultural emissions are estimated at 16.7 kt in 1990, with the majority arising from use of chemical fertilisers (9kt), followed by livestock wastes (5.5kt), biological fixation by leguminous crops (2kt) and crop residue burning (0.2kt). On the same basis, agricultural emissions are expected to fall by 15% by 2000.

[2] *EC Large Combustion Plants Directive.* Directive no. 88/609/EEC.

[3] *Protocol to the 1979 Convention on Long-range Transboundary Air Pollution Concerning the Control of Emissions of Nitrogen Oxides or their Transboundary Fluxes.* UN/ECE, 1988.

[4] *Reducing Emissions of Volatile Organic Compounds and Levels of Ground-level Ozone – A UK Strategy.* Department of the Environment, 1993.

[5] *Protocol to the 1979 Convention on Long-range Transboundary Air Pollution Concerning the Control of Emissions of Volative Organic Compounds or their Transboundary Fluxes.* UN/ECE, 1991.

CHAPTER 7

ASSISTANCE TO DEVELOPING COUNTRIES AND COUNTRIES WITH ECONOMIES IN TRANSITION

DEVELOPING COUNTRIES

7.1 This Chapter outlines how the UK will meet its Convention commitments (see box) to support developing countries. The Chapter also describes the UK's programme of assistance to the economies in transition in Central and Eastern Europe.

7.2 The UK recognises that, in order for developing countries to progress towards sustainable social and economic development, their energy consumption will need to grow, taking into account the possibilities for greater energy efficiency and for the control of greenhouse gas emissions in general.

7.3 The Convention's objective to stabilise greenhouse gas concentrations can only be realised with the active involvement of developing countries. Some have experienced five fold growth in CO_2 emissions over the last 25 years. IEA has estimated that energy related emissions from developing countries may more than double by 2010, overtaking emissions from OECD countries in the process. The developing countries that ratify the Convention will be committed, amongst other things, under Article 4.1 to prepare inventories of sources and sinks of greenhouse gases, and to formulate and implement programmes of measures to mitigate climate change.

> **UK CONVENTION COMMITMENTS**
>
> The UK, along with other developed countries, will provide new and additional financial resources to meet the incremental costs for developing countries of implementing measures covered by Article 4.1 of the Convention. The Government will also provide new and additional resources to meet the full costs for developing countries of communicating information in accordance with Article 12.

7.4 The UK will also assist developing countries that are especially vulnerable to the adverse effects of climate change with the costs of adaptation to those adverse effects, in accordance with Article 4.4. It will also, in accordance with Article 4.5, provide finance for technology and know-how, to enable other countries to implement the Convention.

7.5 The UK's financial assistance to developing countries will be provided through the financial mechanism of the Convention, and through bilateral, regional and other multilateral channels.

BILATERAL ASSISTANCE

7.6 The UK provides bilateral aid to many developing countries, currently about £670 million each year. The objectives of this aid programme include the promotion of economic reform and long term growth, the reduction of poverty, improved education and health, improved status of women, and assistance with national environmental problems. At the United Nations Conference on Environment and Development (UNCED) in Rio, the Prime Minister announced that substantial extra resources would be directed to support the goals of Agenda 21, including assistance for energy efficiency and forestry conservation.

Energy efficiency

7.7 Energy efficiency brings economic benefits to developing countries whilst reducing the global environmental impact of energy use. It is a very effective way of tackling the issue of increasing developing country greenhouse gas emissions which combines national and global benefits. The UK's aid programme includes a number of energy efficiency initiatives (see box).

> **UK ENERGY EFFICIENCY AID INITIATIVES**
>
> A £90 million project will improve the efficiency of production and transmission of electrical power in Western and Southern India, through improvements to hydropower stations and the construction of a High Voltage DC link. In China, one project in Taiyuan will help to identify ways of reducing environmental damage caused by the energy needs of local industry and housing. Another in Beijing will improve energy efficiency in building. Part of a soft loan agreement with Indonesia worth £64 million will be used to construct a fuel efficient gas-fired combined cycle power station.

7.8 The Government is funding a number of initiatives aimed at rehabilitating energy supply systems and strengthening technical, financial and managerial capabilities, including projects to improve power generation and transmission in India, Pakistan, Bangladesh and Uganda. Wider economic reform programmes are also important to limit greenhouse gas emissions, and the UK funds projects in Pakistan, Ghana and Uganda to address the key issue of energy pricing and to increase the efficiency of fuel use. Several projects also promote renewables, such as hydropower and geothermal energy, and the Renewable Natural Resources Programme includes studies of the environmental effects of hydrodams.

Forestry

7.9 Chapter Four describes Government action to protect forests as carbon sinks in this country. In global terms, however, tropical deforestation is a much bigger problem. Most recent estimates by the Food and Agriculture Organisation of the United Nations suggest that 15.4 million hectares of tropical forest are being lost each year. The main causes of deforestation are conversion of forests to agricultural land, demand for fuel wood and unsustainable logging operations. Other causes include fire, mining, industrial and infrastructural developments, human settlements and overgrazing by livestock. Growing population densities increase these pressures.

7.10 UK aid to the forestry sector primarily aims at the management of forests and trees, for example agroforestry, for the sustainable supply of forest products in their widest sense, including the global benefits of providing carbon sinks. The UK believes the best way of conserving forests in developing countries is to help to maximise the economic and social benefits of forests and forest-based industries in a sustainable way.

> **BILATERAL AID TO FORESTRY**
>
> Bilateral aid to forestry is being focused in three broad areas:
> - institutional development, including policy analysis and planning, to strengthen developing countries' capacities to manage their forests effectively;
> - sustainable forest management and conservation, including the identification of incentives for local people living in and around forests to manage them sustainably; and
> - rural development forestry, including agroforestry, which, when linked to sustainable agriculture, is an important means of helping combat deforestation by stabilising agriculture and producing wood products on farms.

7.11 The UK has increased its aid to the forestry sector significantly in recent years. In 1988/89 the UK spent £7.4 million on bilateral forestry aid. By 1991/92 the figure had increased to about £18 million. In 1992/93 the UK spent £28 million, and now has about 200 forestry projects either underway or in preparation at a total cost to the aid programme of about £150 million.

> **UK FORESTRY AID INITIATIVES**
>
> Bilateral aid to forestry is presently focused in eighteen countries, including Cameroon, Nigeria, India and Brazil. The largest is a £25 million forestry project to help forest conservation in the Western Ghats in India, where the UK is committed to a six-year project to explore with the forest department in Karnataka innovative ways of resolving the various forest users, whilst conserving the forests' ecology. An Indo-British forestry initiative was signed in September 1993, under which the UK and India plan to carry out joint research and studies, and to promote international co-operation on forest issues. The two governments also plan to host jointly an international workshop to prepare for the consideration of forestry by the Commission for Sustainable Development in 1995. A project in Brazil is studying the impact of deforestation on climate in order to improve the accuracy of General Circulation Models of climate change (see Chapter Nine).

Technology co-operation

7.12 One of the most effective ways in which developed countries can help to combat climate change is by facilitating access to technologies which can help limit greenhouse gas emissions. The UK is taking a number of steps to promote, facilitate and finance technology transfer and support endogenous capacities, as required by Article 4.5 of the Convention. The transfer of technology and know-how is a central element of most UK aid projects in developing countries, and makes a vital contribution to the success and sustainability of those projects. Technology and know-how are also transferred through the multilateral assistance programmes funded by the UK. However, by far the largest source of technology transfer to developing countries is through the private sector.

7.13 To help provide a practical solution to the problems associated with commercial technology co-operation and to enable businesses in developing countries to gain better access to environmental technologies, the Prime Minister launched the UK Technology Partnership Initiative in March 1993 (see box).

TECHNOLOGY PARTNERSHIP INITIATIVE

The Initiative aims to improve conditions for increasing successful transfers of environmental technologies to developing countries on a commercial basis. It will do this over three years, primarily through improving direct contacts between businesses in developing countries and the UK through increased dissemination of information on best practice. Technologies covered by the Initiative include those which can be used to increase energy efficiency and promote renewable energy.

The Global Technology Partnership conference in March 1993 was a first step in this Initiative. It brought together over 100 key business representatives from newly industrialising developing countries with a similar number of UK business executives to explore the scope for making technologies available to developing countries. The Conference generated much interest both in the UK and overseas.

The Initiative has five main elements:
- a guide to UK sources of environmental technology, which lists over 400 British companies offering environmental technology;
- bulletins on specific solutions to industrial environmental problems in developing countries;
- information on best environmental practice;
- assistance to UK companies to provide "hands on" training to key business executives from developing countries; and
- a technology co-operation network designed to maximise the flow of information about environmental problems and solutions, including feedback from developing country businesses about local needs and difficulties.

7.14 In developing the Initiative, DTI is also forming closer links with international organisations which have technology programmes, such as the United Nations Environment Programme's (UNEP) International Cleaner Production Information Clearing House. The UK is also continuing to contribute to schemes being run by the EC which facilitate technology transfer, including the European Investment Partners Scheme, which provides financial support for investments in the form of EC and local joint ventures in the developing countries of Asia, Latin America and the Mediterranean.

7.15 The UK, together with eleven other IEA/OECD member countries, is contributing to the financial backing for the initial two year developmental phase of GREENTIE (the Greenhouse Gas Technology Exchange). GREENTIE is being piloted by the IEA and OECD to give interested users access to information on technologies and practices which help reduce greenhouse gas emissions, and particularly to help developing countries to locate information on greenhouse gas abatement technologies.

Other bilateral assistance

7.16 The Government is also providing a range of other bilateral assistance to developing countries, which will meet the UK's commitments under the Convention, including:
- technical advice to demonstrate the benefits of energy efficiency in developing countries (see paragraphs 7.7 to 7.8);
- advice to encourage better policy analysis of forestry programmes (see paragraphs 7.9 to 7.11);
- provision of training places for officials from developing countries on the Technical Co-operation Training Programme; and
- assistance to the Environmental Protection Councils in Ghana and Nigeria.

7.17 The UK considers it essential that developing countries are fully involved in the IPCC process, and in negotiations for the implementation of the Convention, and has funded the participation of developing countries in the IPCC and the INC. The Government has contributed £70,000 over the last two years to the IPCC Trust Fund which enables developing country participants to attend IPCC meetings, and will support the participation of lead authors from developing countries in the preparation of the IPCC Second Assessment Report. The UK has also funded a series of seminars on climate change for Ministers and senior officials from developing countries : the "IPCC Roadshow". The UK has also contributed £150,000 to the INC's developing countries Trust Fund over the last two years.

7.18 The UK aid programme also seeks to involve the private and voluntary sectors wherever possible. The Government works with the private sector through the

Commonwealth Development Corporation and Joint Venture Initiatives. Non-Governmental Organisations have been involved through the Joint Funding Scheme and the consultation process on the Sustainable Development Strategy, and 17 UK NGOs and over 60 developing country NGOs attended the "Partnerships for Change" Conference in Manchester in 1993.

MULTILATERAL ASSISTANCE

Global Environment Facility

7.19 Article 21.3 designates the Global Environment Facility (GEF) of the United Nations Development Programme, the United Nations Environment Programme and the World Bank as the interim financial mechanism of the Convention. The UK has contributed £40.3 million to the GEF's pilot phase. This contribution is made through the Global Environment Assistance Programme of the Government's Overseas Development Administration (ODA). This programme is separate from and additional to the Government's aid programme for developing countries. A wide variety of projects are funded through the GEF.

> **PROJECTS FUNDED THROUGH GEF**
>
> During the pilot phase, a total of 43 projects to address climate change, mainly in Africa, Asia and Latin America, have been approved by the Participants Assembly of the GEF, at a total cost of US$296 million. The GEF has also funded projects in its three other focal areas: biodiversity, international waters, and ozone depletion. Implementation of some of these climate change projects in developing countries has already begun.

7.20 The UK believes that the resources of the GEF must be substantially increased, in order to provide the finance to the developing countries required by the Convention. The UK has called for a $2 to $3 billion replenishment of the core fund of the GEF. This would be three to four times the size of the $800 million core fund committed during the pilot phase. The UK will contribute its share of the replenishment, from the ODA's Global Environmental Assistance Programme, when the size of the replenishment is known.

7.21 The next phase of the GEF is likely to proceed in two stages. During Stage One, from 1994/95, the GEF will concentrate on developing the institutional capacity of developing countries to implement the Convention, and on helping them to prepare the inventories and national strategies required by Article 12. During Stage Two, from 1996 onwards, the GEF will concentrate on supporting the measures taken by developing countries under the Convention to mitigate climate change.

7.22 A valuable method of working with developing countries to formulate national programmes under the Convention is through country studies. The UK funded a number of country studies in 1992. They included studies in Ghana and Kenya on the possible impacts of climate change, and a study in Zimbabwe on options for limiting emissions of greenhouse gases. The UK has also worked with the Organisation of Eastern Caribbean States on the possible impacts of sea level rise.

7.23 Country studies will play an increasingly important role in the implementation of the Convention. The UK believes that to be most effective country studies should, wherever possible, use available IPCC methodologies, and be cost effective, targeted, of value to decision makers and involve local institutions.

7.24 The UK believes that the GEF's role under the Convention should be made permanent when the interim arrangements are reviewed at the first Conference of Parties.

Other multilateral aid programmes

7.25 The UK provides a range of other multilateral assistance for developing countries. The UK makes a major contribution to the aid programmes of United Nations (UN) development agencies and the International Finance Institutions. In 1992/93, the Government contributed £784 million through multilateral channels, and in 1993/94 we expect to contribute £829 million. These figures include UK contributions to the World Bank, the World Bank Group, the European Community and the Regional Development Banks, all of which fund projects in developing countries related to the implementation of the Convention, including forestry conservation.

7.26 The UK participates in the Tropical Forestry Action Plan (TFAP), which aims to encourage donor co-ordination and finance for tropical forests, and the development of integrated planning for forest development at national level. The UK has pressed for the TFAP to become more responsive to developing country needs and to encourage greater participation by local communities and NGOs. The UK is also providing assistance to the International Centre for research in Agroforestry and the Centre for International Forestry Research, and supports the efforts of the International Tropical Timber Organisation to achieve its target for all internationally traded tropical timber to come from sustainably managed sources by 2000.

COUNTRIES WITH ECONOMIES IN TRANSITION

7.27 The Convention recognised in Article 4.6 the unique position of the "economies in transition" (EITs) in Central and Eastern Europe and the former Soviet Union. The UK is committed, along with other countries, to providing multilateral aid through international programmes and bilateral support to help these countries to tackle the particular environmental problems that they face. The EITs contribute around 20% of total world CO_2 emissions, although this share is likely to decrease over the next twenty years.

Multilateral assistance

7.28 The UK participates in multilateral programmes which provide help to Eastern Europe to combat climate change. The UK share of the EC's PHARE programme of assistance to Central and Eastern Europe is around 16% of the programme's 1993 budget of 1.1 billion ecu, which is about £850 million. The programme provides support for economic and social restructuring. It includes both national and regional environmental programmes, which contain elements related to air pollution monitoring and reduction in Bulgaria, Hungary, Poland and Romania. Up to the end of 1992, 240 million ecu of PHARE funds had been allocated to the environmental sector. The UK also strongly supports the European Energy Charter and the negotiations towards a substantial Treaty on energy efficiency and other protocols, which will translate the Charter's principle into legally binding commitments. The European Bank for Reconstruction and Development (EBRD) in London promotes investment to Central and Eastern Europe and the Former Soviet Union (see box).

EUROPEAN BANK FOR RECONSTRUCTION AND DEVELOPMENT

The EBRD provides investment and technical assistance programmes currently worth over ECU 2.3 billion and ECU 109 million respectively. The UK contributes to its capital funds and has provided £1.75 million for its technical assistance activities. The EBRD is required by its charter to promote environmentally sound and sustainable development in the full range of its activities. Projects include help to reduce reliance on polluting heavy industries, to increase the efficiency of energy use and supply, and to promote balanced development of transport systems.

Bilateral assistance

7.29 The UK also provides bilateral aid to Eastern Europe through the Know How Fund (KHF), which administers a £5 million programme, spread over three years, specifically targeted at environmental problems in Eastern and Central Europe, the Former Soviet Union and the Baltic States. The KHF financed a study to assess the production of a greenhouse gas emissions inventory in Romania. It has funded feasibility studies for British companies exploring the possibilities of joint ventures with their Eastern European counterparts in this and many other areas. It has helped to take forward proposals to improve energy efficiency in parts of Russia and to explore the potential for energy efficient gas distribution schemes in the Ukraine. Other initiatives under this programme include an energy efficiency project and public awareness campaigns in Poland and energy and environmental audits in the steel industry in Hungary.

CHAPTER 8

EDUCATION, TRAINING AND PUBLIC AWARENESS

NATIONAL LEVEL

8.1 Improving citizens' awareness and knowledge of climate change issues is essential to enable them to take informed decisions. Many measures aimed at mitigating the adverse effects of climate change, such as saving energy or driving cars more fuel efficiently, depend on effective action by individual citizens. The importance of education is recognised in the Convention. Article 4.1(i) commits the UK "to promote and co-operate in education, training and public awareness related to climate change." Article 6 expands this to call for countries to facilitate education at both national and international levels.

8.2 One of the quickest and most cost-effective ways of reducing CO_2 emissions is by improving energy efficiency and reducing unnecessary waste of energy. Within the UK, awareness programmes for citizens on climate change and ways to mitigate its effects have focused on encouraging changes of attitude and behaviour through information, advice and publicity on energy efficiency. Some £130 million has been spent on energy and fuel efficiency awareness and advice to combat climate change since 1990. This is an integral part of the CO_2 Programme (see Chapter Three) and includes the EEO's advice schemes and publicity programmes, the CO_2 partnership programme, fuel efficiency information, and funding of specific projects by non-governmental organisations.

Participation

8.3 Chapter One describes the Government's wide consultation process on the CO_2 Programme, and Chapter Three sets out how the Government is building a partnership approach to limit CO_2 emissions. Chapter Three gives examples of the action that is being taken by organisations across all sectors to raise awareness and provide advice. A newsletter[1] is being produced bi-annually to provide information on the action being taken by Government and by businesses, the public sector, individual citizens and in the transport sector, to reduce CO_2 emissions. Organisations are encouraged to contribute articles to the newsletter on the action they are taking. Reports of progress of other countries will also be reported, with updates on progress in reducing other greenhouse gases.

Involving Citizens

8.4 Participation is also a central part of the marketing strategy for the "Helping the Earth Begins at Home" publicity campaign. The Government launched the £15 million three year campaign in November 1991 to raise awareness of the link between climate change and energy use in the home, and increase the take-up of energy efficiency measures. It is aimed primarily at home-owners. The campaign runs national television and press advertisements and promotional activities, and provides information on the various energy efficiency measures that can be taken. It emphasises the extent to which such measures can reduce fuel bills and CO_2 emissions.

8.5 Research carried out throughout the period of the advertising campaign demonstrates that it has achieved some success in increasing awareness of the link between climate change and energy use. The use of television has been particularly influential in raising awareness of the issue. Influencing citizens' behaviour is likely to take longer, and the Government will be considering this in its review of the future of the campaign.

INVOLVING CITIZENS IN THE "HELPING THE EARTH BEGINS AT HOME PUBLICITY CAMPAIGN"

Organisations such as local government, voluntary organisations, libraries, manufacturers and retailers, financial institutions, fuel utilities, domestic property agents and the media have all been encouraged to take part in the campaign and help promote its messages to householders. In addition to advertising, "Helping the Earth" Focus Weeks were held in October 1992 and 1993 to enable these third parties to become involved in the campaign through displays, exhibitions and distribution of campaign literature, supported by energy efficiency features and advertising in the local and regional press.

The Government recently set up a "Green Brigade" to encourage children to take practical environmental action. The Green Brigade and the Helping the Earth Begins At Home campaign recently produced an activity pack, "Every Home Helps"[2] to encourage young people to lead the way on saving energy at home.

The Government recognises the opportunities afforded by the media to help promote climate change messages to children. Successful examples include BBC Television's childrens' television programme "Blue Peter" which, in collaboration with the "Helping the Earth Begins At Home" campaign, ran a competition on energy saving lightbulbs in Summer 1993; and a competition organised for children at the Ideal Home Exhibition in March 1993.

8.6 The Government recognises the need for a body to co-ordinate and promote the teaching of energy efficiency in schools. The Government considers that the Centre for Research, Education and Training in Energy (CREATE) is best placed to carry out this role and has recently increased its support to assist with the development of a coherent and integrated approach to the teaching of energy efficiency. In addition, CREATE's private sector funding is also increasing.

8.7 The Government provides a wide range of information aimed at raising awareness about renewable energy including exhibitions and conferences, and issuing a wide range of literature. In addition, the Government is funding the preparation of an Open University Renewable Energy information pack and accompanying videos and is also helping to fund other educational material.

8.8 As part of the work on environmental geography and science, the national curriculum for schools includes items on energy and energy efficiency. This gives teachers an opportunity to introduce ideas on climate change and the effective use of energy. School children can carry out energy audits of their classrooms and their homes (including use of transport). Targets can be set to reduce energy use, and progress monitored by calculating both savings of CO_2 emissions and money. Many voluntary organisations and local authorities are actively engaged in environmental education. The Economic and Social Research Council (ESRC) has commissioned research into the use of computer simulations in the teaching of climate change to primary school children and on the use of multi-media computer technologies to increase children's awareness of these issues.

8.9 In the higher education sector, many universities have introduced material on climate change issues into relevant undergraduate and masters courses. Support for postgraduate training is provided by many research councils through both Master of Science and Doctor of Philosophy schemes to ensure there is a cadre of trained manpower to sustain both research programmes and industry.

8.10 DOE encourages environmental and voluntary organisations to promote energy and fuel efficiency to citizens. Some organisations activities are given in Chapter Three. Through its Environmental Action Fund, the Government has contributed £30,000 to Friends of the Earth to enable them to undertake an educational project on air quality, which incorporates energy efficiency.

8.11 The Government also contributes to funding of the Atmospheric Research and Information Centre (ARIC)'s Global Climate Change Information Programme (GCCIP), which is based at Manchester Metropolitan University. The GCCIP was established in October 1991 to provide a public information service on climate change issues. It also aims to make information available to teachers and pupils and to foster awareness of the importance of climate change from an early age. The programme has responded to over 3,500 requests for information since 1991. Teaching packs and education resources are being prepared by the staff and students of the Faculty of Education at the University in liaison with GCCIP. Its services have been advertised to all local education authorities in the UK.

8.12 Agencies of the Inter-Agency Committee on Global Environmental Change will be involved in an event being organised by the UK Global Environment Research (GER) Office as part of the 1994 meeting of the British Association for the Advancement of Science, which will include discussions on climate change.

Involving Businesses and the Public Sector

8.13 The EEO's Best Practice Programme is the Government's main information transfer programme for cost-effective energy efficiency measures. Its aim is to advance and spread the efficiency with which energy is used in the UK in all sectors of industry and commerce, except agriculture and transport. The Best Practice Programme for buildings, including housing, is managed and promoted for the EEO by the Building Research Energy Conservation Support Unit (BRECSU) at the Building Research Establishment. The industry element is managed and promoted for the EEO by the Energy Technology Support Unit (ETSU) at Harwell.

FUEL EFFICIENCY ADVICE

Encouraging citizens to buy more fuel efficient cars and to drive more efficiently is also important to help reduce CO_2 emissions. DOT recently produced a leaflet with information and advice – "Motoring and the Environment"[3]. The "New Car Fuel Consumption" guide[4] provides detailed fuel efficiency information and is regularly updated. The Government is setting up a forum with motor manufacturers and others, to co-ordinate publicity on fuel efficiency issues.

ENERGY DESIGN ADVICE SCHEME

The Government's Energy Design Advice Scheme (EDAS) opens a channel of communication between those involved in design decisions and experts on low energy technologies applied to buildings and raises awareness and understanding of energy issues throughout the construction industry.

8.14 Under Best Practice, the EEO produces a range of publications providing independent, authoritative guidance on energy efficiency. Best Practice information is made available via published case studies, guides, workshops and seminars, some in collaboration with professional bodies, energy suppliers and trade associations.

8.15 The EEO also produces, as part of the Best Practice Programme for school buildings, information and guidance to help school headteachers, governors and caretakers improve energy efficiency. Although the aim is to improve the energy efficiency of the buildings, this helps to raise awareness generally about energy efficiency as part of the curriculum. In a similar vein, seminars and training packages aimed at industrial, commercial and public sectors generally inform audiences about opportunities at home as well as at work.

8.16 DOE publishes information on the availability of training courses on energy efficiency and assists organisations of all sizes with help and advice on energy efficiency training. The Department is considering the formation of a body with a leading interest in developing qualifications and standards for training in energy efficiency.

8.17 The new Energy Saving Trust (see Chapter Three) is overseeing a three year pilot trial of local energy advice centres, partly funded by the Government. The Government is supporting ten "green" business clubs, in partnership with the private sector, to help small businesses improve their environmental performance, including their energy efficiency.

8.18 Advice to farmers on ways to minimise agricultural emissions of greenhouse gases is provided in the MAFF/Welsh Office Code of Good Agricultural Practice for the Protection of Air[5]. MAFF also sponsors its technical agency (ADAS) to run an energy efficiency programme to offer advice and guidance to farmers and horticulturalists.

Involving Scientists and Technologists

8.19 The scientific challenge to understand and predict climate change and its impacts, and to develop technical and socio-economic options to avert or adapt to climate change, is a long-term challenge. It will demand a flow of trained specialists in a wide range of disciplines, not simply to do more research but to work within industry, commerce and education to develop and implement solutions. Many Universities have already introduced climate change issues into undergraduate and masters degree courses and the Research Councils are supporting training schemes in relevant disciplines.

8.20 The Natural Environmental Research Council (NERC) supports advanced courses in applied climatology and meteorology, atmospheric science, remote sensing and climate modelling while the ESRC is extending its programme of research studentships on the socio-economic aspects of global environmental change. The special problems of teaching such a wide-ranging subject, at every educational level, are being studied by the Education and Global Change Project based at the University of York, and by ESRC commissioned research on methods for teaching global change issues from primary school level.

INTERNATIONAL LEVEL

8.21 The UK is committed to co-operation and promotion of climate change material at the international level. Chapter Seven gives more details of the education and training programmes which assist developing countries and the newly independent states of central and eastern Europe and the former Soviet Union.

8.22 The UK has taken the opportunity of leading Working Group I, Science Assessment of the Intergovernmental Panel on Climate Change, to promote wider understanding of the science of climate change. Under the chairmanship of Sir John Houghton, supported by a Government-funded technical secretariat at the Hadley Centre, the Working Group's first assessment report of 1990 and the Supplementary Report of 1992 (see Chapter One) have been prepared as basic resource texts and published rapidly by Cambridge University Press to ensure the widest possible availability. The Government intends that this pattern of rapid, open publication will continue with future reports.

> **HELPING DEVELOPING COUNTRIES**
>
> Following publication of the IPCC first Science Assessment Report, the Government helped fund a series of seminars in developing countries about climate change and the conclusions of the Report. This included a secondment to the IPCC Geneva secretariat to establish and organise the programme.

8.23 Non-Governmental bodies are also assisting with the exchange of information. ARIC's GCCIP information service is actively seeking to establish an international network for information on climate change. As a first step, GCCIP is seeking international agreements which allow information to be shared and disseminated between organisations within the European Community and Eastern Europe. Discussions are taking place within the United Nations Environment Programme's Information Unit on Climate Change in Geneva to negotiate collaborative work on a European Climate Change Information Network and to explore the connections that can be made with the Government's LINK project. GCCIP is also providing researchers and information to developing countries. A network of contacts in developing countries is being developed and GCCIP is preparing a book on "North-South Perspectives on Climate Change".

NOTES

[1] *"Climate Change" newsletter.* The Department of the Environment, first edition Autumn 1993 (published biannually).

[2] *Every Home Helps.* The Department of the Environment, 1993.

[3] *Motoring and the Environment* leaflet. The Department of Transport, 1993.

[4] *New Car Fuel Consumption* booklet. The Department of Transport (published biannually).

[5] *Code of Good Agricultural Practice for the Protection of Air.* The Ministry of Agriculture, Fisheries and Food/Welsh Office Agriculture Department, 1992.

CHAPTER 9

UK CLIMATE CHANGE RESEARCH

UK CONTRIBUTION TO INTERNATIONAL RESEARCH

9.1 The UK is committed under Article 4.1g of the Convention to promote and co-operate in research into climate change to reduce the uncertainties regarding its causes, effects, magnitude and timing. The focus of international assessment for climate change science is the IPCC established in November 1988. The UK has led the Science Assessment, Working Group I, since the inception of IPCC, through the chairmanship of Sir John Houghton (who under the latest revised IPCC procedures becomes co-chair with Brazil). The Government's funding of a technical secretariat unit for Working Group I at the Hadley Centre has been a crucial factor in ensuring that the science assessments have been comprehensive, authoritative, peer-reviewed and promptly published.

9.2 The Government is also supporting contributions by scientists, economists and other specialists to the IPCC's work on assessing climate change impacts, and adaptation and response strategies. The Government will continue to support this work and will remain a leading financial contributor to IPCC as it prepares its Interim Report in 1994 for the first Conference of Parties to the Convention and a second full Assessment Report in 1995.

9.3 Research on the oceanic, atmospheric and terrestrial systems that together determine the global climate necessarily involves international co-operation to mount experiments on a global scale. UK scientists have played an important role in the genesis, planning and direction of many of these programmes which are supported on a bilateral or multilateral basis by the UK Research Councils and the Meteorological Office, and through UK contributions to the parent international bodies including the World Meteorological Organisation, the United Nations Environment Programme, the Intergovernmental Oceanographic Commission and the International Council of Scientific Unions. UK contributions include hosting international focal points for the World Ocean Circulation Experiment and the International Geosphere Biosphere Programme (IGBP), providing the secretariat for the proposal for a Global Ocean Observing System, chairing the scientific committee of the IGBP and the Joint Scientific and Technical Committee for the Global Climate Observing System, and membership of the standing committee of the Human Dimensions Programme.

THE UK RESEARCH PROGRAMME

9.4 The UK's national research effort places it at the forefront of climate change research. This effort is summarised in the recent report prepared by the UK Global Environment Research (GER) Office on behalf of the Inter-Agency Committee on Global Environmental Change (IACGEC) (see paragraph 9.5). The IACGEC provides a discussion forum and also acts as a mechanism for the UK funding agencies to co-ordinate and plan the UK approach to global environmental change research, including climate change. The GER Office provides the Secretariat for the IACGEC and acts as a UK focal point for information flow and exchange on national and international science and policy developments.

9.5 The GER Office recently published the second report of the IACGEC, setting out the framework for UK global environmental change research, including climate change. Entitled "Global Environmental Change: the UK Research Framework 1993"[1], the Report provides a context for co-ordination of research across the wide range of subjects and institutions involved, identifies some long-term priorities, and provides a compendium of UK research funding agency responsibilities. The growth in climate change-related research is a major factor in the growing budget for environmental change research (see box).

UK SPENDING ON CLIMATE CHANGE RESEARCH

The IACGEC report shows the total UK agency expenditure on processes, impacts and response measures research for global environmental change rising by over 50% in the last 4 years.

£m

1989/90	1990/91	1991/92	1992/93
90.51	121.82	114.86	140.87

A recent estimate of UK spending on climate change research, which included additional data acquisition programmes, such as the meteorological and oceanographic data used in climate research, puts the total close to £200M in 1992/93.

Data for Climate Change Research

9.6 A critical factor for progress in climate change research is the availability of data to all researchers covering the spectrum from Earth observation satellite data to socio-economic data. The UK is playing a strong role in a number of fora where better data access and distribution are being developed. The IACGEC has recognised the importance of improvements in data management. A first practical step has been the establishment of the UK Global Environmental Network for Information Exchange (GENIE) as the node of a distributed network of UK natural and social science global environmental change data holdings.

9.7 The IACGEC continues to seek ways of ensuring the maximum availability at minimum cost of data to bona fide researchers at national and international levels. The UK has strongly supported the coordinating and planning activities of the Committee on Earth Observation Satellites (CEOS) in the field of international data networking. In 1992 the UK, as chair of CEOS, ensured that a principal theme was identification of the data needs of the major international research programmes relating to climate change and its impacts. At the European level, the UK is contributing to the development of the proposal by the EC for a Centre for Earth Observation, aimed at establishing a distributed network of data sources, processors and users for Europe interfacing with other world data networks.

Climate prediction and monitoring

9.8 The NERC, the Meteorological Office and independent research groups such as the Climatic Research Unit at the University of East Anglia, have established over the last two decades leading international roles in climate process science, modelling of the oceans and atmosphere, and interpretation of climate records. One major contribution has been in the re-analysis of past meteorological data to generate a reliable time series of global average temperature for the past 130 years, which has provided the direct evidence of global warming in the 20th century. DOE and the Meteorological Office built on this strong base of scientific expertise to establish the Hadley Centre for Climate Prediction and Research as one of the world centres for predictive climate modelling (see box).

9.9 Such models are essential to understanding the complex causes, nature and effects of climate change. Their successful development and use depend on utilising the latest information from research on the processes that influence climate, such as ocean circulation, cloud physics, the chemistry of the atmosphere and land processes, in all of which the UK has strength of international stature.

9.10 The British National Space Centre co-ordinates the contributions of the UK agencies supporting earth observation space missions and instruments, including the UK's contribution to the European Space Agency (ESA), which are becoming increasingly important to climate change

---

THE HADLEY CENTRE FOR CLIMATE PREDICTION AND RESEARCH

Set up in 1990, the Hadley Centre is one of the few centres in the world able to operate fully ocean/atmosphere coupled models with transiently increasing atmospheric concentrations of CO_2. It combines long running Meteorological Office Research on climate modelling, funded by the Government at about £3 to £4 million a year, with a Department of the Environment programme of £8 to £9 million per year, to develop and run the most advanced Atmosphere Ocean Coupled General Circulation Models (AOCGCMs) of the world's atmosphere and oceans. The Hadley Centre collaborates with a wide range of research bodies in many of its research projects.

Excellent progress has been made in increasing the validity and reducing uncertainties in climate change prediction including:
- the first transient climate change experiment using an AOCGCM, one of only four such model simulations that have been performed to date; and
- significant progress to reduce uncertainties in areas including the detailed description of oceans, the radiation budget and clouds.

Future experiments at the Hadley Centre are planned to coincide with key UK and international policy requirements, in particular the IPCC assessments. Major areas of future work include:
- extending the transient run model simulation experiments to include past climate as well as a number of future scenarios;
- development of a high resolution model to improve climate prediction at a regional and national level; and
- progress in the detection of climate change, through data assimilation and satellite measurements.

---

monitoring and research. The Science and Engineering Research Council (SERC), through the Rutherford Appleton Laboratory, has supported and led the development of instruments, both for ESA and NASA missions, and means for handling and disseminating data relevant to climate monitoring and process research. Following the recent White Paper on Science, Engineering and Technology[2], responsibility for Earth observation will transfer to the NERC. Recent important missions and achievements include:

CLIMATE CHANGE RESEARCH OF THE NERC

The NERC has the responsibility for deploying Government funding to support high quality scientific research in the environmental sciences through its Institutes, and through support of scientists in Universities. About a third of the NERC budget can be attributed to research relevant to climate change. Significant progress includes:

- climate process studies in the polar regions, through the British Antarctic Survey and Universities, include studies of potential ice-sheet changes, the analysis of ice cores for information about paleo-climatic variability, and studies of impacts on polar ecosystems;
- the Terrestrial Initiative on Global Environmental Research (TIGER) is a multidisciplinary programme on climate processes and impacts involving land surface, including terrestrial elements of the carbon cycle (which affects the rate of build up of CO_2 in the atmosphere), and the land aspects of the energy and water cycles (which are important for assessing both climate change impacts and the feedback to climate of changes in land biota);
- marine science, including the Joint Global Ocean Flux Study, which contributes to understanding the role of the ocean as a sink for CO_2, and the World Ocean Circulation Experiment (WOCE), is essential for building a more realistic representation of the role of the oceans in climate models. The James Rennell Centre at Southampton has been established to support all aspects of the WOCE programme from field studies to analysis of results of modelling;
- the programme of Land Ocean Interaction Studies which aims to understand the processes at the land/ocean interface as a basis for impact research;
- the Universities Global Atmosphere Modelling Programme (UGAMP), which is a consortium of university groups developing models and analysis techniques complementary to the climate prediction modelling at the Hadley Centre;
- development of the use of satellite data to support process and impact studies, and long-term environmental monitoring including, with other agencies, the Environmental Change Network; and
- paleo-climate research, aimed at identifying past climate change, both to improve the study of climate change processes and as an indication of possible future change.

- successful launch and operation of the ESA ERS-I satellite;
- establishment of a continuous high precision and accuracy sea-surface temperature record from the Along Track Scanning Radiometer instrument on ERS-I. This instrument was developed by a consortium led by the Rutherford Appleton Laboratory and the unique temperature record it produces offers a promise of early detection of climate change, as well as contributing to climate process studies and the operation of climate prediction models;
- the UK developed Advanced Microwave Instrument providing wind strength and direction sea-surface data, as well as high resolution surface imagery independent of cloud cover;
- the Improved Stratospheric and Mesospheric Sounder (ISAMS) on the NASA upper atmosphere research satellite, producing atmospheric profiles of chemistry, water vapour and aerosols; and
- the operation of the geophysical data facility for archiving and disseminating earth observation data.

9.11 The UK's prime objective in European Space Agency programmes is earth observation, and this is reflected in the UK's strong support for missions to follow ERS-I. The programme to launch a follow-on satellite, ERS-II, in 1994/95 will include a further development of the Along Track Scanning Radiometer UK instrument, extending measurements to the monitoring of global vegetation. In 1998/99 the ENVISAT-I satellite is proposed to extend further the range of instruments and earth observation measurements and, together with planning for a METOP-I satellite in 2000, will take the earth observation programme for climate research well into the next century.

9.12 An international meeting convened by the Prime Minister in London in April 1992 on better use of earth observation satellite data identified the need to facilitate the provision of information on satellite missions and data to national and international environmental programmes. As a result, the UK prepared the first complete dossier on existing and planned satellite earth observation missions "The Relevance of Satellite Missions to the Study of the Global Environment"[3], and transmitted it to the UNCED conference on Environment and Development in Rio de Janeiro in June 1992. This dossier will now be updated annually through the Committee on Earth Observation Satellites.

9.13 Through the SERC, the Government funds national supercomputing facilities to support programmes of the Research Councils. In the report "Research Requirements for High Performance Computing" published by the SERC, climate change and global modelling were identified as of particular importance when considering further enhancement of facilities. Since mid-1992, the power of the facilities has been increased by a factor of three by installing new computers at the Atlas Centre of the Rutherford Appleton Laboratory and the University of Manchester Computing Centre and by upgrading the computer at the University of London Computer Centre. The capability at the Rutherford Appleton Laboratory is a joint modelling facility, which is the key element in the ocean, atmosphere and land modelling programmes of the NERC.

9.14 The UK is also carrying out major research to improve inventories of greenhouse gas emissions (see Chapters Three, Five and Six) and sinks (see Chapter Four).

Response measures

9.15 Many of the research programmes aimed at identifying or evaluating measures to help prevent climate change are mentioned in other chapters of this report (see Chapters Three, Four, Five and Ten in particular). A wide range of agencies is involved in research projects in this area. These include several Government departments, the SERC, the Building Research Establishment (BRE), ETSU and universities.

9.16 Among the most important areas of work are:
- DTI's Energy and Environment Programme, a series of studies to identify and evaluate the possible role of new energy technologies in greenhouse gas abatement;
- the Energy-Related Environment Issues Programme of the DOE and BRE, whose long-term strategy includes research on energy efficiency technology and standards in building design, lighting, heating and air-conditioning;
- research projects supported by both the DTI and DOT on new and advanced technologies for vehicles and vehicle fuels;
- research conducted by both the Forestry Commission and the Institute for Terrestrial Ecology on carbon sequestration by sinks;
- the DTI's programme to stimulate the development and deployment of renewable energy technologies (see Chapters Three and Ten);
- work commissioned by MAFF's Flood and Coastal Defence Division on assessing response strategies to accelerated sea level rise;
- research sponsored by DTI and DOE to quantify the likely impact on greenhouse gas emissions of different response options; and
- work by the Centre for Social and Economic Research on the Global Environment (CSERGE) on the economic costs of controlling greenhouse gas emissions, and on various socio-economic aspects of climate change.

Impacts and adaptation

9.17 Article 4.1e of the Convention commits countries to co-operate in preparing for adaptation to the impacts of climate change. The Impacts Working Group of the IPCC concluded that, although there are substantial uncertainties, global warming could have important detrimental effects on agriculture, forestry, natural ecosystems, water resources, human settlements, and coastal protection. The priority aim must be to reduce the uncertainties about the impacts of climate change so that the necessary adaptation measures can be properly targeted.

9.18 The Government funds major research projects to assess the possible impacts of climate change in the UK, in particular through the NERC and the Agriculture and Food Research Council (AFRC). These include studies on possible impacts on agriculture, fisheries, forestry, water resources and ecology. The work of CSERGE is also complemented by the major ESRC Global Environmental Change research programme, which focuses on the socio-economic dimensions of climate change. The programme includes a wide range of economic modelling and assessment projects, the impact of climate change on land use, and verification and implementation of international agreements. A report produced by the UK Climate Change Impacts Review Group[4] in 1991 concluded that climate change in the UK could have serious implications for water resources, soil moisture (affecting, for example, land use and the stability of buildings) and wildlife. MAFF is also funding work aimed at assessing the potential impacts of climate change, both in terms of crop growth, pest and disease incidence, and land use in the UK, as well as the opportunities and challenges which might arise as a result of changes in agricultural markets elsewhere in the world.

9.19 Sea level rise would put additional pressure on coastal defences which have to be taken into account in design and management decisions, particulary in vulnerable areas such as the East Coast, the Thames Estuary and Lancashire. MAFF is conducting on-going research into the possible impacts of accelerated sea level rise.

MAFF RESEARCH INTO IMPACTS OF CLIMATE CHANGE

MAFF has sponsored research to assess the long term economic impact of climate change in two case study areas, one of which has been developed into a vulnerability assessment using the IPCC common methodology. Other work has investigated the impact of climate change on beaches, and there are major on-going studies to improve long-term coastal modelling techniques which can be used to investigate the effects of climate change scenarios at a much greater level of detail.

MAFF also sponsors long-term sea level measurement around the UK. Current work on the use of global positioning systems using satellite ranging techniques for the establishment of absolute levels at tide guage sites is showing promising results, with repeatability at the sub centimetre range. This will be important for detecting real changes in sea level.

9.20 A Government report published in September 1993[5] assessed the impacts of the exceptionally prolonged warm spell in the UK between 1988 and 1990, with temperatures 2°C above normal and low rainfall in the south and east. This was not, in itself, evidence of global warming, but it could certainly be regarded as an illustration of one of the possible impacts of climate change. The report provided a wide-ranging account of the impacts of this short-term change in the normal climatic pattern on plants and animals in the natural environment, of the response of different ecosystems, and of impacts on agriculture, horticulture and forestry. Examples of the impacts included a tendency of plants to flower and animals to breed early, increased insect activity and good yields of vegetables and winter cereals. The report will help guide further UK research into impacts of climate change.

NOTES

[1] *Global Environmental Change: the UK Research Framework 1993.* UK Global Environmental Research (GER) Office, 1993.

[2] *Realising Our Potential – A Strategy for Science, Engineering and Technology.* HMSO, 1993. Cm 2250. ISBN 0-10-122502-4.

[3] *The Relevance of Satellite Missions to the Study of the Global Environment* (UK Dossier). Committee on Earth Observation Satellites, 1992.

[4] *The Potential Effects of Climate Change in the United Kingdom.* UK Climate Impacts Review Group. HMSO, 1991.

[5] *Impacts of the Mild Winters and Hot Summers in the United Kingdom in 1988-90.* Cannell, MGR, and Pitcairn, CER (eds). HMSO, 1993. ISBN 0-11-752642-8.

CHAPTER 10

BEYOND 2000: TOWARDS SUSTAINABILITY

> **OBJECTIVE OF CONVENTION**
>
> The ultimate objective of this Convention and any related legal instruments that the Conference of the Parties may adopt is to achieve stabilisation of greenhouse gas concentrations at a level that would prevent dangerous anthropogenic interference with the climate system. Such a level should be achieved within a time frame sufficient to allow ecosystems to adapt naturally to climate change, to ensure that food production is not threatened and to enable economic development to proceed in a sustainable manner.
>
> Article 2, Framework Convention on Climate Change

CLIMATE CHANGE BEYOND 2000

10.1 The measures contained in the preceding chapters of this Report are intended to meet the commitments in the Framework Convention on Climate Change as it currently stands, with the aim in particular of returning emissions of greenhouse gases to 1990 levels by the year 2000. But we cannot ignore the question of what may need to be done beyond 2000. The Convention itself requires that the adequacy of those commitments be reviewed by the first meeting of the Conference of Parties – likely to take place in early 1995 – and again by 1998.

10.2 The IPPC is continuing its work to increase understanding of the science of climate change. This will enable us to predict with greater certainty the consequences of continuing increases in global greenhouse gas emissions, including the climatic effects and economic impacts which might result. That work will feed into the review of the Convention's existing commitments. It will inform decisions about whether, and if so how, to turn it into something more than a framework Convention, and what measures may be needed beyond 2000. The UK intends to play a full part in a science-based re-appraisal of the Convention's requirements.

10.3 The Convention's ultimate objective, as set out in Article 2, provides the basis for a sustainable approach to climate change. Defining the level at which atmospheric concentrations of greenhouse gases must be stabilised in order to avoid dangerous interference in the climate system will be a major scientific and political task. Agreeing internationally on how to achieve it in the face of ever-increasing population and economic growth (especially in the developing world) could well prove to be one of the major pre-occupations of the twenty-first century.

10.4 Although there is still at present no unequivocal evidence of the impact of greenhouse gas emissions, the advice from the IPCC suggests there is a strong and growing probability that human-induced climate change is taking place, and a significant risk of serious impacts resulting. If the scientific evidence continues to point in this direction, it seems certain that the existing commitments in the Convention would need to be extended and strengthened if its objective is to be achieved. New approaches may need to be developed to help meet such commitments, including comprehensive targets covering emissions of all greenhouse gases, and arrangements to enable Parties to implement them jointly in ways which maximise the extent and the cost-effectiveness of the savings that can be achieved.

THE UK CONTRIBUTION

10.5 As the previous Chapter indicates, the UK is making a major contribution to the scientific effort to improve the prediction of climate change and its effects. The Government's continuing commitment to support the efforts of the IPCC, in particular through the Hadley Centre for Climate Prediction and Research (see Chapter Nine), is a reflection of the importance which we attach to this work.

10.6 With the measures in this programme in place, UK emissions of most greenhouse gases should decline beyond 2000. However, this may not be the case for the most important gas, CO_2. As Figure 10a shows, the range of scenarios in EP59 indicates that CO_2 emissions could rise steadily beyond 2000 (although of course the further into the future they project, the greater the degree of uncertainty becomes). Even with the measures in this programme in

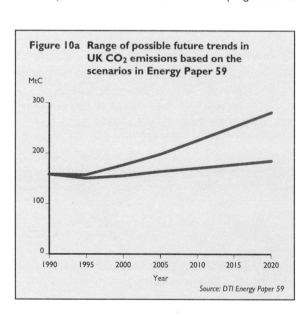

Figure 10a Range of possible future trends in UK CO_2 emissions based on the scenarios in Energy Paper 59

Source: DTI Energy Paper 59

place, on the basis of the scenarios, it is likely that UK emissions of CO_2 would increase again beyond 2000.

10.7 If there is scientific and political consensus on the need for new commitments beyond 2000, the UK, along with other countries, will need to consider what further measures can be taken to limit emissions of CO_2 in particular. Since some of those measures could well involve a long lead-time before results are achieved, and may entail longer term changes to the way in which our society and our economy operate, it is important that we begin at an early stage to consider the possibilities. The remainder of this Chapter looks at possible options for action which will have a significant impact beyond 2000 in four important areas.

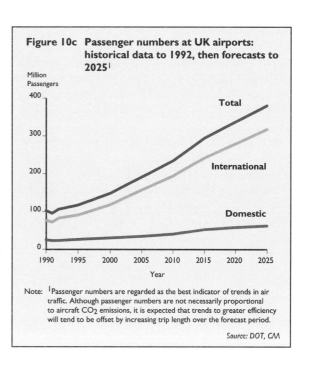

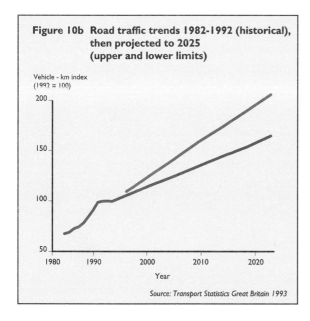

Transport

10.8 In the Energy Paper 59 scenarios, CO_2 emissions from transport continue to increase beyond 2000 The current National Road Traffic Forecasts show traffic continuing to rise at a steady rate through to the end of the current projections in 2025 (see Figure 10b). By then traffic would be between 65% and 105% higher than in 1989. In 2025 car ownership would still not have reached the theoretical ceiling where all those who are able to drive cars have them, so further strong growth could be projected beyond the 2025 date. Air traffic is set to rise at even more rapid levels into the next century (see Figure 10c), though the treatment of CO_2 emissions from aviation is still to be agreed internationally. These strong growth trends will continue to put pressure on the ability of the UK to curb levels of CO_2 emissions (see Figure 10d). They also pose significant threats to other aspects of the environment.

10.9 The Government will be looking at ways of reconciling these pressures with the constraint of sustainable development. This is likely to involve measures to take better account of the environmental costs of transport and influence the rate of traffic growth. The overall goal will be to meet the needs of the economy for access to markets in ways which substantially reduce the need to travel and the impacts of that travel.

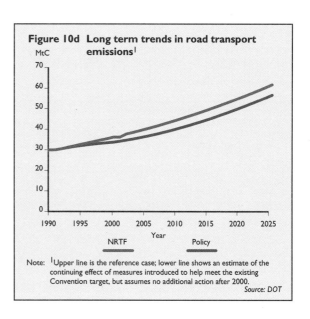

10.10 Measures to be considered might include:
- broad fiscal measures to increase the cost of fuels and increase incentives for efficiency in the choice and use of vehicles (building on the existing commitment to increase vehicle fuel duties by on average at least 5% per year above inflation);
- specific market measures, such as congestion charging, that will increase the marginal cost of transport to the user and reflect wider costs to others, so increasing overall efficiency;
- continuing policies to develop land in ways which reduce the need to travel and hence emissions (draft guidance to local authorities on using their planning powers to achieve these aims has already been issued);
- working with the motor industry to achieve maximum improvements in vehicle fuel efficiency (estimates suggest improvements of up to 40% are within reach over the next decade or so, through a combination of improved technology and decisions by motorists to buy more efficient models);
- work, especially by local authorities, to develop integrated packages of transport and land use policies which maximise the potential for walking, cycling and efficient use of public transport;
- maximising the economic transport of goods by rail and sea, including development of new rail freight terminals and further examination of the potential role of rail freight and the barriers to its use in the UK;
- investment in transport infrastructure where this contributes both to achieving the access needed for the economic health of the country and to meeting environmental objectives;
- regulatory measures to increase vehicle efficiency, either directly through standards (set at European Community level), or indirectly, for example through reduced speed limits or improved vehicle testing;
- development of standards and market measures in international fora, including on emissions from marine and air transport; and
- generating public awareness about the implications and environmental costs of the growth in transport emissions and the need to modify behaviour.

10.11 More research is needed to help evaluate these options. Tackling transport pressures effectively will require a better understanding of a range of key issues. Given the vital economic role of transport, it will be important to be able to assess the cost-effectiveness of each option, both in its own right, and against the impact of other options and measures, in other sectors. Knowing more about the factors which influence people's decisions on transport and location will help the formulation of local policies to reduce the need to travel. Research developments in telecommunications and information technology will also clarify the potential effects of new developments in these fields on transport and location issues.

10.12 Further work on the environmental and congestion costs of transport should help us to quantify the many hidden costs of transport use, and so enable prices to be set at a more economically efficient level, to the benefit of the environment. Study of the interaction between transport provision and its use should inform future national policy development, as will work to distinguish between scheme-level impacts and wider environmental concerns. The Government proposes to commission a comparative study on the effects of different transport fuels, including potential new fuels such as compressed natural gas, liquified petroleum gas, biofuels and electricity.

10.13 The Government will be exploring these issues as part of its further work on sustainable development. Several are already the subject of study. These results will inform policy development into the next decade.

10.14 The public will have a major role to play in meeting the challenge of growing transport emissions. Transport trends are the sum total of all our decisions as individuals. To have any major impact on projected trends, measures would be required that would change current patterns of transport use. People may not always be able to make journeys as easily or as cheaply as before. Costs of travel, particularly road travel, may well need to rise to reflect environmental costs and constrain demand. Individuals will need to reconcile their desire for travel with their desire to protect the environment for future generations.

Electricity Generation

10.15 In the EP 59 scenarios, electricity generation is the fastest growing source of CO_2 emissions beyond 2000. This is partly due to rising demand for electricity as it continues to account for an increasing proportion of energy use. It is partly also the result of a reduction in the share of generating capacity accounted for by nuclear, since it is assumed that as the advanced gas cooled reactor (AGR) and Magnox plants reach the end of their active lives they are not replaced by new nuclear capacity (this assumption has been made without prejudice to the outcome of the forthcoming review of nuclear power in the UK (see paragraph 10.18)). Projected emissions from fuel use in other sectors – excluding transport – remain broadly stable (and they may in fact fall given the continuing impact beyond 2000 of measures to improve energy efficiency taken as part of the programme set out in Chapter Three). As already mentioned, there is considerable uncertainty about possible trends so far ahead, but if CO_2 emissions from electricity were to rise substantially in the first two decades of the next century as indicated in the EP 59 scenarios, this would pose a considerable challenge for the limitation of UK emissions.

10.16 Limiting emissions from electricity generation can be achieved by improving the efficiency of end use to reduce electricity demand (see paragraphs 10.22 to 10.24), by improving the efficiency of generation or by changing the balance of fuels used in generation.

10.17 Competition in the electricity generating market will help to ensure that generation is carried out as efficiently as possible. The contribution of combined heat and power schemes is expected to continue to grow. Around 20% of capacity might be met from CHP schemes by 2020. This could bring substantial further reductions.

10.18 The Government has announced its intention to review the future prospects for nuclear power in the UK. As can be seen from paragraph 10.15, decisions on the future strategy for nuclear power in the UK will have a significant impact on CO_2 emissions. For example, a nuclear power station like Sizewell B has a generating capacity of around 1200MW. Figure 10e illustrates the approximate amount of CO_2 that would be produced in generating an equivalent amount of electricity from other possible sources.

Figure 10e	Illustrative carbon emissions from 1200 MW of fossil fuel capacity at 80% load factor
Fuel	**MtC/yr**
Coal	2
Oil	1.6
Gas (CCGT)	0.9
	Source: DTI

10.19 Renewable energy currently provides around 2% of the current UK electricity supply. As described in Chapter Three, the Government's policy is to stimulate the development of new and renewable energy sources wherever they have prospects of being economically attractive and environmentally acceptable. In the short term, the most promising technologies for the UK in this respect are solar, wind, hydro and wastes; and in the longer term, energy crops, fuel cells and photovoltaics.

However, the geothermal technologies, tidal, offshore wind and wave, seem unlikely to contribute substantially to UK energy supply in the foreseeable future. The Government's strategy for stimulating the development of renewable energy technologies has recently been reviewed. The revised strategy will be published early in 1994.

10.20 The future contribution of renewable energy sources towards UK energy supply depends upon the success of current programmes and relative energy costs. By 2025, renewables may be supplying between 15 and 60 TWh/y on a competitive basis, equivalent to 5 to 20% of current electricity supply, with further contributions from the heat producing renewables such as passive solar design and biofuels.

10.21 Notwithstanding the contributions which might come from renewable and nuclear, the likelihood is that fossil fuels will continue to be the major contributor to electricity supply well into the next century. The Government will continue to monitor the market to examine whether the balance of energy sources is compatible with meeting the UK's environmental obligations. If action is necessary to alter the balance, market-based instruments, including fiscal measures, would be likely to be the preferred option.

Energy Consumption

10.22 The measures in the CO_2 programme described in Chapter Three will reinforce the underlying trend towards improvements in energy efficiency in the UK. Overall, energy efficiency improvements equivalent to some 12 to 15% of total consumption of energy outside the transport sector in 1990 are expected to be achieved by the year 2000. Beyond 2000 there is likely to remain considerable potential for cost-effective improvements in energy efficiency to be achieved. Perhaps up to 25% of current total energy consumption outside the transport sector could be saved through take up of energy efficiency investments with a pay back of up to 5 years, and if all buildings and plant were replaced (as they come up for replacement) with the best economic energy efficiency techniques and technology available at the time. However, experience has shown that decision makers do not necessarily adopt what appear to be economic energy efficiency measures and that considerable effort, or strengthening of the financial incentives to energy efficiency, is needed to secure much of the available savings.

10.23 The Government will be working to ensure that as much as possible of the potential savings are achieved, taking forward the programme now in place. Standards for the energy efficiency of buildings and equipment will be kept under review and strengthened as technology improves. New and improved energy efficiency standards for products are likely to be agreed as the European Community's SAVE programme is developed.

10.24 More generally, the operation of the energy supply markets will be monitored to examine whether barriers exist to achieving efficiency in energy use. Since without external stimulus, voluntary action to improve energy efficiency tends to be limited, past experience suggests that higher energy prices may be needed if the UK were to take on tighter international commitments on greenhouse emissions in the future.

Afforestation

10.25 Chapter Four describes the potentially important role of forests as sinks for absorbing carbon. This is recognised both in the Convention and in the Statement of Forest Principles that was adopted at the Earth Summit in Rio. Measures to protect and restore the earth's forests could play an increasingly important role in future actions under the Convention.

10.26 The creation of new woodlands and forests could in the future make a significant contribution towards the UK's efforts to help limit increases in atmospheric concentrations of CO_2. Although the potential contribution may be small compared to other possible measures, it could play a role in this country's CO_2 programme beyond 2000. Every extra 50,000 hectares of tree planting could absorb around 5 MtC (equivalent to around 3% of the UK's current annual carbon emissions). As set out in the UK's Sustainable Forestry Programme, the Government recognises the role of forests as carbon stores and is committed to a steady expansion of tree cover. Continuation of the current trend towards taking agricultural land out of production under the CAP could provide substantial scope for tree planting.

10.27 When grown as an energy crop, timber can also help to reduce the use of fossil fuels and provide a valuable source of renewable energy that is largely carbon-neutral. The Government is committed to promoting the use of wood for energy and is sponsoring research and the development of pilot projects for harvesting, processing and converting waste and short rotation coppice to energy. The Government is also supporting the establishment of short rotation coppice through its Woodland Grant Scheme.

CONCLUSION

10.28 As we enter the next Century, new and difficult choices may have to be made if we are to achieve a sustainable future. We will need to find ways of breaking the link between economic development and increasing emissions of greenhouse gases, in particular CO_2. New technologies may emerge that will help to alleviate the problem, but they cannot be relied upon to solve it. Unequivocal evidence of the existence of human-induced climate change is unlikely to be available for some years, but meanwhile, taking a precautionary approach, we need to plan ahead now. This Chapter sets out some of the options.

10.29 The adequacy of the existing commitments under the Convention is scheduled to be reviewed by the end of the decade. It is too early to predict what the outcome of this review might be, but it seems likely that countries will need to take on new commitments if the Convention's objective is to be achieved. Substantial emissions reductions would be likely to involve economic costs and would therefore need to be agreed internationally in the light of emerging scientific knowledge. However, the earlier options for action are identified, and the greater the co-operative effort achieved, the less will be the costs.

10.30 The Government will therefore examine these and other options so that the UK is well placed to make a full and appropriate contribution beyond 2000 to achievement of the Convention's ultimate objective. The UK will work hard within the appropriate international fora towards ensuring that objective can be met. Our national programme will be reviewed and developed during the coming years in the context of the international response to the problem of climate change.

ANNEX A

LIST OF ABBREVIATIONS

AA
Automobile Association

ACBE
Advisory Committee on Business and the Environment

ACEA
Association of European Car Manufacturers

ALFED
Aluminium Federation Ltd

AOCGCMS
Atmosphere Ocean Coupled General Circulation Models

ARIC
Atmosphere Research and Information Centre

BATNEEC
Best Available Techniques Not Entailing Excessive Cost

BCC
British Ceramic Confederation

BCEMA
British Combustion Equipment Manufacturers Association

BRECSU
Building Research Energy Conservation Support Unit

BRE
Building Research Establishment

CAA
Civil Aviation Authority

CAP
Common Agricultural Policy

CBI
Confederation of British Industry

CCGT
Gas-fired combined cycle plant

CEOS
Committee on Earth Observation Satellites

CF$_4$
Carbon tetrafluoride

C$_2$F$_6$
Hexafluoroethane

CFCs
Chlorofluorocarbons

CHP
Combined Heat and Power

CHPA
Combined Heat and Power Association

CH$_4$
Methane

CIA
Chemical Industries Association

CLGEF
Central and Local Government Environment Forum

CO$_2$
Carbon dioxide

CREATE
Centre for Research, Education and Training in Energy

CSERGE
Centre for Social and Economic Research on the Global Environment

CTC
Cyclists' Touring Club

DOE
Department of the Environment

DTI
Department of Trade and Industry

DUKES
Digest of UK Energy Statistics

EBRD
European Bank for Reconstruction and Development

EC
European Community

EDAS
Energy Design Advice Scheme

EEO
Energy Efficiency Office

EITs
Economies in transition

EMA
European Eco Management and Audit

EMAS
Energy Management Assistance Scheme

EP59
Energy Paper 59

ESA
European Space Agency

ESI
Electricity Supply Industry

ESRC
Economic and Social Research Council

ETSU
Energy Technology Support Unit

FTA
Freight Transport Association

GCCIP
Global Climate Change Information Programme

GDP
Gross Domestic Product

GEF
Global Environment Facility

GENIE
Global Environmental Network for Information Exchange

GER
Global Environmental Research Office

GREENTIE
Greenhouse Gas Technology Exchange

GW
Gigawatts

HCFCs
Hydrochlorofluorocarbons

HFCs
Hydrofluorocarbons

HMIP
Her Majesty's Inspectorate of Pollution

IACGEC
Inter-Agency Committee on Global Environmental Change

IEA
International Energy Agency

IGBP
International Geosphere Biosphere Programme

INC
International Negotiating Committee on the Climate Change Convention

IOD
Institute of Directors

IPC
Integrated Pollution Control

IPCC
Intergovernmental Panel on Climate Change

ISAMS
Improved Stratospheric and Mesospheric Sounder

KHF
Know How Fund

LCPD
Large Combustion Plants Directive

LRTAP
Long Range Trans-Boundary Air Pollution

MACC
Making a Corporate Commitment

MAFF
Ministry of Agriculture, Fisheries and Food

Mt
Million tonnes

MtC
Million tonnes of carbon

MTOE
Million tonnes of oil equivalent

MW
Megawatts

NERC
Natural Environmental Research Council

NFFO
Non-Fossil Fuel Obligation

NGOs
Non-Governmental Organisations

N_2O
Nitrous Oxide

NO_x
Nitrogen Oxides

NO_2
Nitrogen Dioxide

ODA
Overseas Development Administration

OECD
Organisation for Economic Co-operation and Development

OFFER
Office of Electricity Regulation

OFGAS
The Office of Gas Supply

OPCS
Office of Population Censuses and Surveys

PES
Public Expenditure Survey

PFCs
Perfluorocarbons

PHARE
EC Programme for providing environmental assistance to Eastern Europe

PPG
Planning Policy Guidance

RAC
Royal Automobile Club

RAL
Rutherford Appleton Laboratory

RECs
Regional Electricity Companies

REEOs
Regional Energy Efficiency Offices

SAP
Standard Assessment Procedure

SAVE
Specific Actions for Vigorous Energy Efficiency

SERC
Science and Engineering Research Council

SRO
Scottish Renewables Obligation

tC/ha
Tonnes of carbon per hectare

TFAP
Tropical Forestry Action Plan

TIGER
Terrestrial Initiative on Global Environmental Research

UGAMP
Universities Global Atmosphere Modelling Project

UK
United Kingdom

UK-NAEI
United Kingdom National Atmospheric Emissions Inventory

UKOOA
United Kingdom Offshore Operators Association

UN
United Nations

UNCED
United Nations Conference on Environment and Development

UNECE
United Nations Economic Commission for Europe

UNEP
United Nations Environment Programme

UNGA
United Nations General Assembly

VAT
Value Added Tax

VOCs
Volatile organic compounds

WOCE
World Ocean Circulation Experiment

WSL
Warren Spring Laboratory

ANNEX B

UK NATIONAL ATMOSPHERIC EMISSIONS INVENTORY

B.1 The UK National Atmospheric Emissions Inventory (UK-NAEI) has been published, updated and revised annually since 1987. The UK-NAEI contains emissions time series from 1970, and aims to be as complete and accurate as possible. Details of the methodology were first published in 1987 (see reference 1) and have been kept up to date since (see reference 2).

B.2 The UK-NAEI is used to fulfil the data requirements of the international agreements consolidated under the UNECE Convention on Long Range Trans-Boundary Air Pollution (LRTAP). Therefore the UK-NAEI is already open to scrutiny in an international policy domain. The UK, in conjunction with Germany and latterly the European Community, leads a Task Force to harmonise LRTAP inventory methodologies under the EC CORINAIR system. Development of the IPCC and CORINAIR methodologies is now being co-ordinated at Task Force level.

B.3 Table B.1 and footnotes set out the correspondences between IPCC source categories and UK-NAEI 1990 emissions estimates for CO_2, CH_4, N_2O, NO_x, CO and VOCs. The rest of Annex B outlines the relationship between the UK-NAEI estimation methodologies and those in IPCC's draft guidelines, and gives some indication of the uncertainties involved.

Carbon Dioxide: Fuel and Industrial Processes

B.4 CO_2 emissions from fossil fuel combustion shown in Table B.1 are estimated by sector using data from DUKES (see reference 3) and the emission factors set out in reference 2. For the energy data used in the calculation of emissions, DUKES is compatible with the international energy statistics published by the IEA which underlie the draft revised IPCC methodology. Uncertainties in emission factors are unlikely to exceed a few percent.

Carbon Sinks

B.5 Mapping of the vegetation carbon sink is described in reference 4 and an account of the development of the soils map is in preparation (see reference 5).

B.6 The UK's approach to estimating the effect of land use change on the terrestrial carbon sink is at least equivalent to IPCC's emerging draft methodology, in which changes in carbon sequestered by soil and vegetation are estimated from the product of total area involved in land use change over the appropriate time horizon, and the corresponding time averaged carbon flux per unit area. For afforestation the UK has detailed historical data on new planting, and a full dynamic calculation has been undertaken. An additional sub-category E (Wetland drainage and peat extraction) has been added to Section V of the IPCC summary Table B.1. This reflects the dominant position of peat in the UK terrestrial carbon sink.

B.7 The entries in Section V of Table B.1 are given in more detail in Figure 4d and paragraph 4.9 of the main report. Table B.2 summarises the references and assumptions made in arriving at the estimates in paragraph 4.9. The estimated carbon fluxes in paragraphs 4.10 (semi-natural processes) and 4.19 (straw incorporation and sewage sludge disposal) incorporate advice from the ITE on the relevant carbon dynamics.

Methane

B.8 The UK methane inventory has been revised following assessment by the Watt Committee (see reference 12). The Watt Committee results contributed to the supporting material prepared for IPCC's consideration in the development of draft revised guidelines (see reference 13), and UK procedures are generally comparable with IPCC's more detailed (Tier 2) approaches where these have been drafted. Table B.3 provides a qualitative overview of the procedures used for the main categories of emissions.

B.9 The UK has made estimates of uncertainties in the principal methane source categories. This was achieved by using Monte Carlo analysis to combine expert judgements about uncertainties in emission factor and activity data, and is fully described in reference 14. The best estimate for total emissions in 1990 was about 5 Mt with a range between 3.5 and 7 Mt, and 95% of trials falling between about 4 and 6 Mt.

Nitrous oxide

B.10 Stage I IPCC guidelines are available for the major sources of nitrous oxide (see reference 15), and preliminary methodologies for the UK nitrous oxide inventory have been outlined (see reference 2). The estimates in Table B.1 draw on these sources. Stage I IPCC guidelines are not available for nitrous oxide emissions from animal wastes and estimates for this source were based on the latest research findings from MAFF. The considerable uncertainties involved in estimating emissions of nitrous oxide have not been quantified to date. Table B.4 provides a qualitative overview of the procedure used to estimate emissions.

Other Greenhouse Gases

B.11 Detailed national emissions of nitrogen oxides, volatile organic compounds and carbon monoxide have been published by source category in the UK-NAEI (see reference 2). As mentioned in paragraph B.2, these data are submitted to meet UK requirements under international agreements and the methodologies are open to extensive scrutiny. Quantitative details of the methodologies are given in the relevant source documents quoted in the UK-NAEI.

References

1) Eggleston H S and McInnes G: *Methods for the Compilation of UK Air Pollutant Emission Inventories*; Report LR 634 (AP), Warren Spring Laboratory, 1987.

2) Gillham C A, Leech P K and Eggleston H S: *UK Emissions of Air Pollutants 1970–1990*; Report LR 887 (AP), Warren Spring Laboratory, 1992; see also references therein. Sixth annual report (covering period 1970-1991) is in press.

3) Department of Trade and Industry: *Digest of United Kingdom Energy Statistics*; published by the Government Statistical Service, HMSO, London (annual).

4) *The Carbon Content of Vegetation, and its Geographical Distribution in Great Britain:* Milne R; Institute of Terrestrial Ecology, Edinburgh Research Station, 1993.

5) *The Carbon Content of Soil, and its Geographical Distribution across Great Britain:* Howard P, et al, Institute of Terrestrial Ecology, Merlewood Research Stations (in preparation).

6) *Agricultural Census:* UK Ministry of Agriculture, Fisheries and Food, London (annual).

7) Cannell, M G R and Milne R: UK Institute of Terrestrial Ecology, assessment of available data.

8) Cannell M G R and Dewar R C: The Carbon Sink Provided by Plantation Forests and their Products in Britain paper in preparation.

9) Assessment based on range in Cannell M G R, Dewar R C and Pyatt D G: Conifer Plantations on Drained Peatlands in Britain: a Net Gain or Loss of Carbon? *Forestry*, 66 (4), 353–369, 1993.

10) Loveland P J: UK Soil Survey and Land Research Centre, Cranfield, assessment of unpublished data.

11) Howard P: UK Institute of Terrestrial Ecology, assessment of available data.

12) *UK Methane Emissions:* Watt Committee on Energy, London, in press.

13) A R van Amstel (ed): *Proceedings of the International Workshop on Methane and Nitrous Oxide*, Amersfoort, The Netherlands, February 1993; RIVM 1993.

14) Bellingham J R, Milton M J T, Woods, P T, Passant N R, Poll A J, Couling S, Marlowe I T, Woodfield M, Garland J and Lee D S: *The UK Methane Emissions Inventory: a Scoping Study on the use of Ambient Techniques to Reduce Uncertainties*. Report QU98, UK National Physical Laboratory, 1993.

15) *Estimation of Greenhouse Gas Emissions and Sinks*: Final report of the OECD Expert's Meeting, Paris February 1991. Prepared for IPCC, OECD Paris, 1991.

Table B.1. UK emissions for 1990 by IPCC source category

Source categories/emissions	CO_2 / kt of C	CO_2 / kt	CH_4 / kt	N_2O / kt	NO_x / kt, a	CO / kt	VOC / kt, b
Total National Emission	158255	580268	4844	109, j	2779	6701	2692, m
I. All energy (fuel combustion and production, transmission, storage and distribution)							
A. Fuel combustion activities							
energy and transformation industries	62669	229745	5	p	835	57	14
industry (ISIC)	25869	94851	7	p	183	63	2
transport	32913, c	120681, c	11	8	1559, c	6065, c	1142, c
commercial/institutional	8296	30413	2	p	56	8	1
residential	21797	79922	49	p	68	286	40
agriculture/forestry	733	2688	0.2	p	4	1	0.1
other	d	d	(3)	3, k	d	d	d
biomass burning for energy	(e)	(e)	(e)	(e)	(e)	(e)	(e)
B. Fuel production, transmission, storage and distribution							
crude oil and natural gas	1545, r	5665, r	481	(e)	51	1	308
coal mining			756				(e)
II. Other industrial production processes (ISIC)							
A. Chemicals			(5)	80	9		295
B. Non-metallic mineral products	2024	7421					
C. Other (ISIC)	2409, f	8833, f	0.5		12	220	38
III. Solvent and other product use			(e)	(e)	(e)	(e)	752
IV. Agriculture							
A. Enteric fermentation			1077				
B. Animal wastes			485, q	14, l			
C. Rice cultivation			na	na			
D. Agricultural soils			h	4, l			
E. Agricultural waste burning			(18)	0.2			
F. Savannah burning	na	na	na	na	na	na	na
V. Land-use change							
A. Forest clearing	na	na	na	na	na	na	
B. Conversion of grasslands to cultivated lands	(0 ± 500, g)	(0 ± 1883, g)					
C. Logging/Managed forests	(-2500)	(-9167)					80, m
D. Abandonment of managed lands	(e)	(e)					
E. Wetland drainage and peat extraction	(500)	(1833)					
VI. Public service							
A. Landfills			1900				19
B. Sewage treatment			71, i				
C. Other	n	n					n

Key to Table B.1

Numbers in columns may not add up due to rounding errors.

( ) Entries in parenthesis means that the source category does not contribute to the National Emissions Total at the top of the table.
na Not applicable to UK
a Expressed as NO_2 equivalents
b Excluding methane
c Includes emissions due to aircraft ground movements and landing and take-off cycle up to 1km and from UK shipping in coastal waters (<12 miles). For carbon dioxide, these emissions are 713 ktC and 1759 ktC respectively.
d Included under commercial/institutional
e Not estimated, but thought to be small
f Incineration and landfill gas flaring
g See paragraph B.6 and 4.10 in the main Report
h Agricultural soils could be a net sink of methane
i Latest estimates from the Water Services Association suggest that methane emissions from the disposal of sewage sludge may be overestimated by up to 20%.
j UK-NAEI estimate for current nitrous oxide emissions of 175 kt includes emissions from all soils. No time series is given. See also note l.
k All fuel combustion except vehicles
l New IPCC background material (see reference 13) applied to the UK gives 5.5 kt for nitrous oxide from animal wastes, 9 kt from fertilizer application to agricultural soils, and 2 kt from biological fixation of nitrogen by agricultural crops and grassland. UK-NAEI unpublished revised estimate is 40 kt total flux of N_2O from all UK soils.
m Total includes natural emissions from forest growth of 80 kt
n Included elsewere
p Included under *other* fuel combustion
q New IPCC background material (see reference 13) gives 108 kt for UK methane emissions from animal wastes and 24 kt from agricultural waste burning.
r Includes emissions from gas flaring but excludes other emissions from offshore platforms.

Table B.2 Summary of assumptions on carbon sinks.

Entry in Fig 4.d	Summary of Assumptions
Agricultural land use change (0±0.5 MtC/yr)	Estimate based on long term trends in land use (see reference 6), together with an assessment of associated carbon dynamics (see reference 7).
Afforestation (+2.5 MtC/yr)	Dynamic model of Carbon stored in trees, litter and products (see reference 8), using Sitka Spruce as the representative species and driven by annual statistics on new planting back to 1925. This approach is considerably more detailed than the IPCC methodology.
Drainage of deep peat (-0.2 MtC/yr)	190 000 ha total drained area, assumed to lose a representative 100 $gC\ m^{-2}\ yr^{-1}$ (cf (see reference 9)).
Drainage of lowland wetlands (-0.2 MtC/yr)	70 000 ha drained area, assumed to lose 250 $gC\ m^{-2}\ yr^{-1}$ (see reference 10).
Peat extraction (-0.1 MtC/yr)	In-situ 1500 000 m^{-3} extracted annually with bulk density (dry weight per unit in-situ volume) 110 $kg\ m^{-3}$ and 50% carbon content. Two year half life for assumed for oxidation (see reference 7) of carbon in peat extracted (see reference 11).

Table B.3 Overview of methane inventory assumptions

Source Category	Emission factors/ main methodology	Activity data
Fuel Combustion	Emission factors depend on carbon content of fuels	National energy statistics
Oil and gas exploration and production	Generic installation based emission factors	Statutory reporting of gas flared; census of installations
Coal production	Emission factors from measurements at UK mines	Annual production data for deep and shallow mines
Gas distribution	Statistically stratified leakage tests on national distribution network	Distribution network characteristics; national energy statistics
Other industry	Technology specific emission factors	National energy statistics
Enteric fermentation	Emission factors for representative animal types	Annual animal census
Agricultural wastes	Emission factors by disposal practice; waste arisings by representative animal type	Annual animal census
Landfills	First order decay model calibrated to UK measured data	Population and national data on waste characteristics
Sewage	Emission factors for specific disposal routes	Population and national data on sewage disposal

Table B.4 Overview of nitrous oxide inventory assumptions

Source Category	Emission factors/ main methodology	Activity data
Industry	Emission factor calculated from an assessment of the chemistry involved	Annual production statistics and IPC regulations
Fertiliser use	Emission factors for representative UK fertiliser types	Survey of Fertiliser Practice and sales of fertiliser
Animal wastes	Estimate of extent of denitrification of mineral nitrogen	Annual animal census
Agricultural crop burning	Emission factor based on chemical characteristics of a representative crop	National survey data (crop burning banned from 1993)
Vehicles	Emission factors for general types of vehicles	National transport statistics
Fuel combustion	Emission factors from UK measurements	National energy statistics